Number 313

Michael E. Taylor

Noncommutative microlocal analysis, Part I

Memoirs
of the American Mathematical Society

Providence · Rhode Island · USA

November 1984 · Volume 52 · Number 313 (end of volume) · ISSN 0065-9266

color code: 182
lettering: black

Apbk

Memoirs of the American Mathematical Society
Number 313

Michael E. Taylor

Noncommutative microlocal analysis, Part I

Published by the
AMERICAN MATHEMATICAL SOCIETY
Providence, Rhode Island, USA

November 1984 · Volume 52 · Number 313 (end of volume)

MEMOIRS of the American Mathematical Society (ISSN 0065-9266) is published bimonthly (each volume consisting usually of more than one number) by the American Mathematical Society at 201 Charles Street, Providence, Rhode Island 02904. Second Class postage paid at Providence, Rhode Island 02940. Postmaster: Send address changes to Memoirs of the American Mathematical Society, American Mathematical Society, P. O. Box 6248, Providence, RI 02940.

The paper used in this journal is acid-free and falls within the guidelines established to ensure permanence and durability.

Table of Contents

ABSTRACT: Pseudodifferential operators on $\mathbb{R}^n$ in $OPS^m_{\rho,0}$ are built out of smooth families of convolution operators on $\mathbb{R}^n$. Similarly important classes of operators can be built out of smooth families of convolution operators on a noncommutative Lie group G. When the representation theory and harmonic analysis on G are well understood, one can construct a noncommutative symbol calculus. This paper develops some aspects of the resulting noncommutative microlocal analysis. Chapter I treats operators on general Lie groups. The details of the symbol calculus depend on the particular representation theory of the group G, and such a theory is worked out for the Heisenberg group in Chapter II. In Chapter III this theory is applied to a systematic study of operator classes on contact manifolds, including parametrices for naturally occurring subelliptic operators, heat asymptotics, and a study of the Szegö projectors.

AMS (MOS) subject classificasion (1970): 35H05

Library of Congress Cataloging in Publication Data

Taylor, Michael Eugene, 1946-
Noncommutative microlocal analysis.

(Memoirs of the American Mathematical Society, ISSN 0065-9266 ; no. 313-)
"November 1984, volume 52, number 313."
Bibliography: p.
Includes index.
1. Differential equations, Hypoelliptic.
2. Pseudodifferential operators. 3. Lie groups.
I. Title. II. Series: Memoirs of the American Mathematical Society ; no. 313, etc.
QA3.A57 no. 313, etc. 510 s [515.3'53] 84-18500
[QA377]
ISBN 0-8218-2314-0

NONCOMMUTATIVE MICROLOCAL ANALYSIS, PART I

Introduction. The theory of pseudodifferential operators has provided a very powerful and flexible tool for treating problems in linear partial differential equations. For example, it provides a framework for "pasting together" parametrices for "frozen" constant coefficient elliptic operators (obtained by Fourier analysis on $\mathbb{R}^n$) to produce parametrices for general variable coefficient elliptic equations. Using the theory of Fourier integral operators, one can impose a translation symmetry on operators with simple characteristics and real principal symbol and obtain rich information on such subjects as strictly hyperbolic equations.

In dealing with operators with multiple characteristics, in cases where the geometry of the characteristics can be put in standard form via a homogeneous canonical transformation, one is often led naturally to a problem in noncommutative harmonic analysis. The connection between the Kohn Laplacian and analysis on the Heisenberg group was exploited by Folland and Stein [F4]. More general hypoelliptic operators were studied via analysis on more general nilpotent Lie groups by Rothschild and Stein [R5]. These papers and numerous subsequent ones developed operator calculi for certain classes of pseudodifferential operators. However, they do not exploit harmonic analysis on a nilpotent Lie group in the same way that the most popular approach to pseudodifferential operators on $\mathbb{R}^n$ does, by the group Fourier transform and associated symbol calculus. Some aspects of such a symbol calculus were presented by Dynin in [D2] and [D3], which gave considerable inspiration to this paper.

The sort of operators we deal with arise on a Lie group G from a smooth

Received by the editor Aug. 15, 1983.
Research partially supported by NSF grant MCS8201766A01.

family of convolution operators on G in the same fashion that classical pseudodifferential operators on $\mathbb{R}^n$ arise from a smooth family of Fourier multipliers (convolution operators) on $\mathbb{R}^n$. In the first section of Chapter I, we develop a general study of such operators and the calculus: adjoints, products, pseudo-locality, etc. This theory proceeds along the lines of what is now a standard treatment of pseudodifferential operators on $\mathbb{R}^n$ with symbols in $S^m_{\rho,0}(\delta = 0)$. The symbol of such an operator is defined on $G \times \hat{G}$, where $\hat{G}$ consists of the set of equivalence classes of irreducible unitary representations of G, taking values as operators on the associated representation space. The first chapter presents much of its results in a general framework, though it also focuses on some phenomena special to nilpotent Lie groups with dilations, and occasionally specializes to two step nilpotent Lie groups.

The second chapter develops the tools to implement symbol calculi on the Heisenberg group $\mathbb{H}^n$. Our goal here is to develop harmonic analysis on the Heisenberg group far enough to construct suitably powerful classes of right invariant pseudodifferential operators on $\mathbb{H}^n$ and understand them through their symbols. We describe the basic representations of the Heisenberg group and analyze the image of a convolution operator under such a representation as an operator in the Weyl functional calculus, which has reached a high degree of development in [H10]. Other approaches to harmonic analysis on $\mathbb{H}^n$ have been pursued, particularly by Geller [G2], [G3],[G5],[C3], and others. A more elementary treatment of harmonic analysis on $\mathbb{H}^n$, having some overlap with the one here, is given in the first chapter of the monograph [T5]. We develop a class of pseudodifferential operators on $\mathbb{H}^n$ containing both the convolutors on $\mathbb{H}^n$ with the dilation properties stressed by Folland and Stein and the algebra OPS^0 of classical pseudodifferential operators. Thus there is available classical microlocalization via elements of the algebra.

The methods developed in Chapter I allow one to apply the theory of

Chapter II to the study of classes of pseudodifferential operators on a contact manifold M, including parametrices for hypoelliptic pseudodifferential operators on M doubly characteristic on a certain symplectic variety $\Lambda \subset T^*M \setminus 0$ (defined by the contact structure), satisfying the condition of subellipticity with loss of one derivative, given originally by Sjöstrand [S6], Boutet de Monvel and Treves [B13], and Boutet de Monvel [B7]. The approach here is in straightforward analogy with the construction, via symbol calculus for OPS^m, of parametrices for elliptic operators. Our noncommutative symbols are operator valued, a property in common with part of Sjöstrand's construction in [S6]. In our case, it is apparent that the parametrices obtained are pseudodifferential operators of type $(\frac{1}{2},\frac{1}{2})$; in fact, they are obtained in operator classes which are strict sub-classes of Boutet de Monvel's operator classes in [B7]. Next we construct a parametrix for the "heat" equation determined by such a subelliptic operator, in the case when it is a selfadjoint semibounded second order differential operator, in direct analogy to the construction of a parametrix for the heat equation on a Riemannian manifold. Parametrices have been constructed in greater generality by Menikoff and Sjöstrand [M8], and subsequent papers, using harder work; the approach here is simple enough to proceed from the principal term in the trace of the heat kernel to show the trace of the heat kernel has a complete asymptotic expansion. Finally, we show that the symbol calculus developed in Chapter III gives a very straightforward construction of the Szegö projector, for the boundary of a strictly pseudoconvex domain in $\mathbb{C}^n$. The construction of Boutet de Monvel and Sjöstrand [B12] lends itself to greater generality, but the treatment via noncommutative symbol calculus seems very natural in this context. The fact that Fourier integral operators are avoided has some advantages, e.g., to L^p estimates.

On the other hand, Fourier integral operators greatly add to the flexibility of pseudodifferential operators as a tool, and will probably play a more crucial role in exploiting harmonic analysis on other groups, in situations modeling other types of symplectic geometry than just a contact line bundle in T^*M. It was originally my intention to include more chapters, discussing applications of harmonic analysis on certain other classes of groups, but the present manuscript has grown to a point where a break is necessary. We will not go into the contents of the projected second part, beyond a few references in the text to a fourth chapter. Let us point out the works of Melin [M4] and of Beals and Greiner [B6] which develop certain operator calculi on spaces more general than contact manifolds. The operator algebras which are treated are not amalgamated with the classical pseudodifferential operators, so further work would be required to microlocalize them.

This paper ends with two appendices. The first gives a brief discussion of some aspects of the Weyl calculus. The second develops a more general sort of Weyl calculus and applies it to a proof of a result of Howe [H11] on an isomorphism between a certain algebra of pseudodifferential operators on $\mathbb{R}^n$, which also arises naturally in Chapter II, and an algebra of Toeplitz operators on the unit ball in $\mathbb{C}^n$.

Some of the material of this paper appeared in the informally circulated notes [T4], which by now have been cited several times.

Chapter I: Noncommutative Approach to Pseudodifferential Operators

Here we produce various classes of pseudodifferential operators on Lie groups in a fashion parallel to the way pseudodifferential operators with symbols in $S^m_{\rho,0}$ are produced from convolution operators on $\mathbb{R}^n$. A great deal of the resulting operator calculus arises smoothly in close analogy to the development of pseudodifferential operators on $\mathbb{R}^n$, for very general classes of Lie groups. The discussion of symbols in this chapter is brief, as this aspect of the study is strongly tied to the particular representation theory of the group at hand.

We will use the following definition of the Euclidean Fourier transform on $\mathbb{R}^n$:

$$\hat{u}(\xi) = (2\pi)^{-n/2}\int u(x)e^{-ix\cdot\xi}dx.$$

Then the Fourier inversion formula is

$$u(x) = (2\pi)^{-n/2}\int\hat{u}(\xi)e^{ix\cdot\xi}d\xi.$$

The author will be found guilty of lapses in the text regarding factors of powers of 2π, which may be omitted from many formulas.

Another convenient symbolism we will use is

$$\langle\xi\rangle = (1+|\xi|^2)^{\frac{1}{2}}$$

for $\xi \in \mathbb{R}^n$.

1. Convolution Operators and Pseudodifferential Operators on Lie Groups.

In this first chapter we show how to build operators out of convolution operators on a Lie group, in a fashion analogous to constructing pseudodifferential operators our of convolution operators (Fourier multipliers) on Euclidean space. For orientation, let us recall the definition of certain classes of pseudodifferential operators on $\mathbb{R}^n$. We set

$$(1.1)\qquad p(x,D)u = (2\pi)^{-n/2}\int p(x,\xi)\hat{u}(\xi)e^{ix\cdot\xi}d\xi .$$

Suppose the amplitude $p(x,\xi)$ belongs to $S^m_{\rho,\delta}$, which is to say

$$(1.2)\qquad |D^\beta_x D^\alpha_\xi p(x,\xi)| \le C_{\alpha\beta}\langle\xi\rangle^{m-\rho|\alpha|+\delta|\beta|} .$$

We say $p(x,D) \in OPS^m_{\rho,\delta}$. In the particular case $\delta = 0$, we can characterize elements of $OPS^m_{\rho,0}$ as follows. The space of Fourier multipliers in $S^m_{\rho,0}$.

$$(1.3)\qquad \begin{aligned} p(D)u &= (2\pi)^{-n/2}\int p(\xi)\hat{u}(\xi)e^{ix\cdot\xi}d\xi \\ &= \hat{p}*u(x) \end{aligned}$$

forms a Frechet space $S^m_{\rho\#}$, with seminorms

$$(1.4)\qquad \sup_\xi \langle\xi\rangle^{-m+\rho|\alpha|}|D^\alpha_\xi p(\xi)| = [p]_{\alpha,m,\rho} .$$

Now we can think of $p(x,\xi) \in S^m_{\rho,0}$ as being a smooth family of elements $S^m_{\rho\#}$:

$$(1.5)\qquad p_x(\xi) = p(x,\xi),$$

and if $P(y)$ is the operator of Fourier multiplication by $p_y(\xi)$, then we have

$$(1.6)\qquad p(x,D)u(x) = P(x)u(x).$$

The operator calculus for such pseudodifferential operators is given as follows. If $p(x,D) \in OPS^m_{\rho,\delta}$ and $q(x,D) \in OPS^\mu_{\rho,\delta}$, then, as long as $0 \le \delta < \rho \le 1$,

$$(1.7)\qquad p(x,D)q(x,D) = r(x,D) \in OPS^{m+\mu}_{\rho,\delta},$$

with

$$(1.8)\qquad r(x,\xi) \sim \sum_{\alpha\ge 0} (i^{|\alpha|}/\alpha!)D^\alpha_\xi p(x,\xi)D^\alpha_x q(x,\xi).$$

Note that the general term in this asymptotic expansion belongs to $S^{m-(\rho-\delta)|\alpha|}_{\rho,\delta}$. This result is given in Hörmander [H6]; see also the books [K10],[T2], [T7], on pseudodifferential operators. The result (1.7) is also valid in case $\rho = \delta < 1$, but in this case (1.8) does not give an asymptotic sum, so we do not have a convenient formula for the principal symbol of a product in this case.

Now let G be a Lie group. Since we intend to work locally, we impose a local coordinate system on a neighborhood of the identity element $e \in G$, such that e is the origin 0; we could use exponential coordinates, identifying a neighborhood of $e \in G$ with a neighborhood of 0 in $\mathfrak{g}$, the Lie algebra of G. The Fourier multipliers (1.3) are replaced by convolution operators on G:

$$(1.9) \qquad Ku(x) = \int_G k(y)u(y^{-1}x)\,dy = \int_G k(xy)u(y^{-1})\,dy = \int_G k(xy^{-1})u(y)\Delta(y^{-1})\,dy$$

where dy stands for Haar measure, $\Delta(y)$ the modular function. We suppose $k \in \mathcal{E}'(G)$ is in some Frechet space $\hat{\mathfrak{X}}$ of distributions which are singular only at the origin. Consequently the Fourier transform $\hat{k}(\xi)$ belongs to $\mathfrak{X}$, a Frechet space contained in $C^\infty(\ ')$, where ' denotes the linear dual of $\mathfrak{g}$. We will suppose that $\mathfrak{X} \subset S^m_{\rho\#}$ for some $m \in \mathbb{R}$, $0 < \rho \leq 1$. We say K belongs to $OP\mathfrak{X}$.

If $\hat{k}(y,\xi) = \int k(y,x)e^{-ix\cdot\xi}dx$ is a smooth function of y with values in $\mathfrak{X}$, for y in a neighborhood of $e \in G$, then K(y), defined by

$$(1.10) \qquad K(y)u(x) = \int_G k(y,xz^{-1})u(z)\Delta(z^{-1})\,dz$$

is a smooth function of y, taking values in the Frechet space $OP\mathfrak{X}$. We then associate the operator

$$(1.11) \qquad \mathfrak{K}u(x) = K(x)u(x),$$

and we say $\mathfrak{K} \in OP\tilde{\mathfrak{X}}$.

One thing we can say about such operators is that they are

pseudodifferential operators in the usual sense. Indeed, we have the following simple result.

Proposition 1.1. If $\mathcal{K} \in OP\tilde{\mathfrak{X}}$ with $\mathfrak{X} \subset S^m_{\rho\#}$, then, modulo a smoothing operator,

$$\mathcal{K}u(x) = \int a(x,y,\xi)e^{i(x-y)\cdot\xi}u(y)\,dy\,d\xi \tag{1.12}$$

for an amplitide in $S^m_{\rho,1-\rho}$, i.e., such that

$$|D_y^\gamma D_x^\beta D_\xi^\alpha a(x,y,\xi)| \le C_{\alpha\beta\gamma}\langle\xi\rangle^{m-\rho|\alpha|+(1-\rho)(|\beta|+|\gamma|)}. \tag{1.13}$$

Proof. We can suppose the group law is given by

$$xy^{-1} = \Phi(x,x-y), \quad \Phi(x,0) = 0 \quad (e=0). \tag{1.14}$$

Consequently, if Haar measure is given by $\mathcal{H}(y)dy$, $\mathcal{H}_1(y) = \Delta(y^{-1})\mathcal{H}(y)$,

$$\begin{aligned}\mathcal{K}u(x) &= \int k(x,xy^{-1})u(y)\mathcal{H}_1(y)dy \\ &= \int u(y)k(x,\Phi(x,x-y)\mathcal{H}_1(y)dy \\ &= \int\int_{R^n} u(y)e^{i\Phi(x,x-y)\cdot\zeta}\hat{k}(x,\zeta)\mathcal{H}_1(y)dy\,d\zeta.\end{aligned} \tag{1.15}$$

Now, by (1.14), we can find a smooth invertible matrix $\Psi(x,y)$ such that, near the diagonal,

$$\Phi(x,x-y)\cdot\zeta = (x-y)\cdot\Psi(x,y)^t\zeta. \tag{1.16}$$

Hence, modulo a smoothing operator,

$$\mathcal{K}u(x) = \int\int e^{i(x-y)\cdot\Psi(x,y)^t\zeta}\hat{k}(x,\zeta)\mathcal{H}_1(y)D(x,y)dy\,d\zeta \tag{1.17}$$

where $D(x,y)$ is supported near $x = y$ and equal to 1 on a small neighborhood of $x = y$. If we make the change of variables.

$$\xi = \Psi(x,y)^t\zeta,$$

then we can write, with $\psi(x,y)$ smooth,

$$\zeta = \psi(x,y)\xi,$$

$$D(x,y)d\zeta = \tilde{D}(x,y)d\xi,$$

so

(1.18) $$\mathcal{K}u(x) = \iint e^{i(x-y)\cdot\xi}\hat{k}(x,\psi(x,y)\xi)\mathcal{H}_1(y)\tilde{D}(x,y)u(y)dy\, d\xi,$$

so we have the form (1.12) with

(1.19) $$a(x,y,\xi) = \hat{k}(x,\psi(x,y)\xi)\mathcal{H}_1(y)\tilde{D}(x,y).$$

It is routine to verify that $\mathcal{K} \subset S^m_{\rho\#}$ yields the estimates (1.13).

In the special case of convolution operators, this result was noted by Strichartz [S10]. If $\rho > \frac{1}{2}$, then the multiple symbol $a(x,y,\xi)$ can be reduced to a simple symbol and we have

(1.20) $$u(x) = \int p(x,\xi)e^{ix\cdot\xi}\hat{u}(\xi)d\xi$$

where

(1.21) $$p(x,\xi) \sim \sum_{\alpha\geq 0} (i^{|\alpha|}/\alpha!)D^\alpha_\xi D^\alpha_y a(x,y,\xi)\Big|_{y=x}.$$

See, e.g., [T 2], Chapter II. In the case $\rho = \frac{1}{2}$, one can still write $\mathcal{K}$ in the form (1.20), as shown in Beals [B2], but (1.21) is no longer asymptotic, and one does not obtain a neat formula for the principal symbol.

In the case when G is a step 2 nilpotent Lie group, we can obtain the form (1.20) in the general case, rather directly, as follows. In exponential coordinates, Haar measure coincides with Lebesque measure, and, via the Campbell-Hausdorff formula, we express the group law as

(1.22) $$xy^{-1} = L_x(x-y),$$

where $L_x = I - \frac{1}{2}\text{ad}\, x$ is a smooth family of linear maps on $\mathfrak{J}$. Thus, in this case

(1.23) $$\begin{aligned}\mathcal{K}u(x) &= \int u(y)k(x,L_x(x-y))dy \\ &= \int \int u(y)e^{-iL_x(x-y)\cdot\xi} k(x,\xi)dy\, d\xi \\ &= \int \hat{k}(x,\tilde{L}_x\xi)e^{-ix\cdot\xi}\hat{u}(\xi)d\xi\end{aligned}$$

where $\tilde{L}_x = (L_x^t)^{-1}$. Thus we directly obtain (1.20), with

(1.24) $$p(x,\xi) = \hat{k}(x,\tilde{L}_x\xi).$$

which gives $\mathcal{K} \in OPS^m_{\rho,1-\rho}$. This argument was given in Nagel and Stein [N3] for the case of convolution operators on the Heisenberg group.

From proposition 1.1 we can conclude, by the Calderon-Vaillancourt theorem, that if $\mathfrak{X} \subset S^0_{\frac{1}{2}\#}$, then $\mathcal{K} \in OP\tilde{\mathfrak{X}}$ is continuous on L^2, and on any Sobolev space H^s. For $\mathfrak{X} \subset S^0_{\rho\#}$ with $0 < \rho < \frac{1}{2}$, application of proposition 1.1 yields only very weak continuity results. In fact, it is apparent from first principles that, given any M, if $N > 0$ is large enough and $\mathfrak{X} \subset S^{-N}_{\rho\#}$, $0 < \rho < \frac{1}{2}$, then any $\mathcal{K} \in OP\tilde{\mathfrak{X}}$ maps H^s to H^{s+M}. Sharper continuity results will be considered in the next section.

A point we wish to emphasize is that, even though proposition 1.1 has some uses (it shows $\mathcal{K}$ is microlocal, for example), if $\rho \leq \frac{1}{2}$ it is not an incisive analysis of $\mathcal{K}$. In particular, formulas for compositions of such operators are not obtained in a usable fashion from proposition 1.1, even in case $\rho = \frac{1}{2}$. We now give an account of compositions of operators of the form (1.11).

Suppose that for $m \in \mathbb{R}$ (or one might restrict attention to $m \in \mathbb{Z}$), $\mathfrak{X}^m$ is a nested family of Frechet spaces ($\mathfrak{X}^m \supset \mathfrak{X}^\mu$ if $m > \mu$) with the following properties:

(1.25) $$\mathfrak{X}^m \subset S^m_{\rho\#} \text{ for some } \rho \in (0,1],\ m \geq 0,$$

(1.26) $$\mathfrak{X}^m \subset S^{m\sigma}_{\rho\#} \text{ if } m < 0, \text{ for some } \sigma \in (0,1],$$

(1.27) $$A \in OP\mathfrak{X}^m,\ B \in OP\mathfrak{X}^\mu \Rightarrow AB \in OP\mathfrak{X}^{m+\mu}, \text{ the product being continuous.}$$

The last hypothesis is on the composition of two convolution operators. Verifying it for particular classes $\mathfrak{X}^m$ is often a problem in harmonic analysis on G, which will be taken up in succeeding chapters. Some special cases are studied at the end of this section. Let us note that all these hypotheses are satisfied in the case $\mathfrak{X}^m = S^m_{\frac{1}{2}\#}$ (even easier, in the cases $\mathfrak{X}^m = S^m_{\rho\#}$, with $\frac{1}{2} < \rho \leq 1$); in all these cases we can take $\sigma = 1$ in (1.26). Now suppose

$$A \in OP\tilde{\mathfrak{X}}^m, \quad B \in OP\tilde{\mathfrak{X}}^\mu. \tag{1.28}$$

Say

$$\mathcal{A}u(x) = A(x)u(x) = \int a(x,xy^{-1})u(y)\mathcal{H}_1(y)dy \tag{1.29}$$

and

$$\mathcal{B}u(x) = B(x)u(x) = \int b(x,xy^{-1})u(y)\mathcal{H}_1(y)dy. \tag{1.30}$$

For the composition, we have

$$\begin{aligned} \mathcal{A}\mathcal{B}u(x) &= \mathcal{A}[\int b(x,xy_1^{-1})u(y_1)\mathcal{H}_1(y_1)dy_1] \\ &= \iint a(x,xy_2^{-1})b(y_2,y_2y_1^{-1})u(y_1)\mathcal{H}_1(y_1)\mathcal{H}_1(y_2)dy_1dy_2. \end{aligned} \tag{1.31}$$

Let us compare this with the operator $\mathfrak{C}$, defined by

$$\mathfrak{C}u(x) = A(x)B(x)u(x). \tag{1.32}$$

By hypothesis (1.27), $C(y) = A(y)B(y)$ is a smooth function of y with values in $OP\mathfrak{X}^{m+\mu}$, so $\mathfrak{C} \in OP\tilde{\mathfrak{X}}^{m+\mu}$. We have the formula

$$\mathfrak{C}u(x) = \iint a(x,xy_2^{-1})b(x,y_2y_1^{-1})u(y_1)\mathcal{H}_1(y_1)\mathcal{H}_1(y_2)dy_1dy_2. \tag{1.33}$$

Note that the only difference between (1.31) and (1.33) is in the first argument of b. This suggests making the expansion

$$b(y_2,z) = \sum_{|\gamma|<N} \varphi_\gamma(xy_2^{-1})b_\gamma(x,z) + \sum_{|\gamma|=N} \varphi_\gamma(xy_2^{-1})r_\gamma(x,y_2,z). \tag{1.34}$$

Here, using exponential coordinates, we take

$$\varphi_\gamma(y) = y^\gamma. \tag{1.35}$$

Thus $b_0(x,z) = b(x,z)$, and each $b_\gamma(x,z)$ is a smooth function of x with values in $\hat{\mathfrak{X}}^\mu$. Furthermore, by Taylor's formula with remainder, $r_\gamma(x,y_2,\cdot)$ is a smooth function of x and y_2 with values in $\hat{\mathfrak{X}}^\mu$. Thus we have

$$(1.36)\qquad \mathcal{AB}u(x) = \sum_{|\gamma|<N} A^{[\gamma]}(x)B_{[\gamma]}(x)u(x) + \mathcal{R}_N u(x)$$

where

$$(1.37)\qquad A^{[\gamma]}(y)u(x) = \int a(y,xy_2^{-1})\varphi_\gamma(xy_2^{-1})u(y_2)\mathcal{H}_1(y_2)dy_2,$$

$$(1.38)\qquad B_{[\gamma]}(y)u(x) = \int b_\gamma(y,xy_1^{-1})u(y_1)\mathcal{H}_1(y_1)dy_1,$$

and $\mathcal{R}_N$ is given by

$$(1.39)\ \mathcal{R}_N u(x) = \sum_{|\gamma|=N} \iint a(x,xy_2^{-1})\varphi_\gamma(xy_2^{-1})r_\gamma(x,y_2,y_2y_1^{-1})u(y_1)\mathcal{H}_1(y_1)\mathcal{H}_1(y_2)dy_1dy_2.$$

Suppose $u \in H^s$ is compactly supported. Then crude considerations give $M = M(\mu)$, independent of γ, such that $\int r_\gamma(x,y_2,y_2y_1^{-1})u(y_1)\mathcal{H}_1(y_1)dy_1 = v(x,y_2)$ is a smooth function of x taking values in the Sobolev space H^{s-M}, as a function of y_2. It will be convenient to switch the order here; pick K large and regard v as an H^{s-M} function of y_2, taking values in the Sobolev space H^K (functions of x). Thus $\mathcal{R}_N u(x) = w_N(x,x)$, where

$$\begin{aligned}(1.40)\qquad w_N(x,y) &= \sum_{|\gamma|=N} \int a(x,xy_2^{-1})\varphi_\gamma(xy_2^{-1})v(y,y_2)\mathcal{H}_1(y_2)dy_2\\ &= \sum_{|\gamma|=N} A^{[\gamma]}(x)v(y,x)\\ &= \sum_{|\gamma|=N} \mathcal{A}^{[\gamma]}v(y,x).\end{aligned}$$

Here the operators $\mathcal{A}^{[\gamma]}$ is applied to the H^K valued function v (of x). For N large, the integral operator $\mathcal{A}^{[\gamma]}$ has a very weak singularity, so $w_N(x,y)$ will be in H^K for any given K in N is sufficiently large. Consequently we can interpret (1.36) as an asymptotic expansion

$$(1.41)\qquad \mathcal{AB}u(x) \sim \sum_{\gamma\geq 0} A^{[\gamma]}(x)B_{[\gamma]}(x)u(x).$$

As already noted, $B_{[\gamma]}(x)$ is a smooth function of x with values in $OP\mathfrak{X}^\mu$. As for $A^{[\gamma]}(x)$, in view of (1.37) and (1.35), if $\mathfrak{X}^m \subset S^\nu_{\rho\#}$ (with $\nu = m$ for $m \geq 0$, $m\sigma$ for $m < 0$), then $A^{[\gamma]}(x)$ is a smooth function of x with values in $S^{\nu-\rho|\gamma|}_{\rho\#}$. Thus the terms in (1.40) become highly smoothing for large $|\gamma|$. To obtain results within the framework of $OP\tilde{\mathfrak{X}}^*$, we add the following hypotheses to (1.25)-(1.27):

(1.42) If $p(\xi) \in \mathfrak{X}^m$, then $D_\xi^\alpha p(\xi) \in \mathfrak{X}^{m-\tau|\alpha|}$, for some $\tau \in (0,1]$,

(1.43) If $K_j \in \mathfrak{X}^{m-\tau j}$, then there exists $K \in \mathfrak{X}^m$ such that, for any M, if N is sufficiently large, $K - (K_0+\cdots+K_N) \in S_{\rho\#}^{-M}$.

It follows from these hypotheses and from (1.27) that the γth term in (1.41) belongs to $OP\tilde{\mathfrak{X}}^{m+\mu-\tau|\gamma|}$. We summarize the result we have obtained:

<u>Proposition 1.2</u>. Let $\mathfrak{X}^m$ be a nested family of Frechet spaces satisfying the hypothesis (1.25)-(1.27) and (1.42)-(1.43). If $\mathcal{A} \in OP\tilde{\mathfrak{X}}^m$, $\mathcal{B} \in OP\tilde{\mathfrak{X}}^\mu$, then $\mathcal{AB} \in OP\tilde{\mathfrak{X}}^{m+\mu}$, and, with $\mathfrak{C} \in OP\tilde{\mathfrak{X}}^{m+\mu}$ given by (1.33), we have $\mathcal{AB}-\mathfrak{C} \in OP\tilde{\mathfrak{X}}^{m+\mu-\tau}$. More precisely, the asymptotic expansion (1.41) holds.

We turn to the analysis of adjoints of elements of $OP\tilde{\mathfrak{X}}^m$. If $\mathcal{K}$ is given by (1.10),(1.11), we have

(1.44) $$\mathcal{K}^*v(x) = \int k^\#(y,xy^{-1})v(y)\mathcal{H}_1(y)dy,$$

where

$$k^\#(y,z) = \overline{k(y,z^{-1})}.$$

We compare this with $\mathcal{K}^\#$, defined by

(1.45) $$\mathcal{K}^\#v(x) = \int k^\#(x,xy^{-1})v(y)\mathcal{H}_1(y)dy.$$

We see that $\mathcal{K}^\# \in OP\tilde{\mathfrak{X}}^m$ provided $\mathfrak{X}^m$ satisfies the condition

(1.46) $$p(\xi) \in \mathfrak{X}^m \Rightarrow \overline{p(\xi)} \in \mathfrak{X}^m.$$

The formula (1.45) differs from (1.44) only in the first argument of $k^\#$, so, in analogy with (1.34), we make the expansion

(1.47) $$k^\#(y,z) = \sum_{|\gamma|<N} \varphi_\gamma(xy^{-1})k_\gamma^\#(x,z) + \sum_{|\gamma|=N} \varphi_\gamma(xy^{-1})r_\gamma^\#(x,y,z).$$

In this case, $r_\gamma^\#(x,y,z)$ is a smooth function of x and y, taking values in $\hat{\mathfrak{X}}^m$.

It follows that

$$(1.48) \qquad \mathcal{K}^* v(x) = \sum_{|\gamma|<N} K_{\{\gamma\}}(x)v(x) + \mathcal{R}_N^{\#} v(x)$$

where

$$(1.49) \qquad K_{\{\gamma\}}(z)v(x) = \int k_\gamma^{\#}(z,xy^{-1})\varphi_\gamma(xy^{-1})v(y)\mathcal{H}_1(y)dy$$

and $\mathcal{R}_N^{\#}$ is given by

$$(1.50) \qquad \mathcal{R}_N^{\#} v(x) = \sum_{|\gamma|=N} \int r_\gamma^{\#}(x,y,xy^{-1})\varphi_\gamma(xy^{-1})v(y)\mathcal{H}_1(y)dy.$$

We see that $k^{\#}(x,z)$ is a smooth function of x taking values in $\hat{\mathfrak{X}}^m$, and hence, granted hypothesis (1.42), the map $\mathcal{K}_{\{\gamma\}}$ given by $\mathcal{K}_{\{\gamma\}}v(x) = K_{\{\gamma\}}(x)v(x)$ belongs to $OP\tilde{\mathfrak{X}}^{m-\tau|\gamma|}$. Meanwhile, an analysis very like that of $\mathcal{R}_N$ shows that $\mathcal{R}_N^{\#}$ is arbitrarily smoothing if N is sufficiently large. We have established the following result.

Proposition 1.3. Let $\mathfrak{X}^m$ be a nested family Frechet spaces satisfying the hypotheses (1.25)-(1.27),(1.42)-(1.43), and (1.46). If $\mathcal{K} \in OP\tilde{\mathfrak{X}}^m$, then the adjoint $\mathcal{K}^*$ belongs to $OP\tilde{\mathfrak{X}}^m$, and one has the asymptotic expansion

$$(1.51) \qquad \mathcal{K}^* v(x) \sim \sum_{\gamma \geq 0} K_{\{\gamma\}}(x)v(x).$$

As we have noted, the hypotheses giving propositions 1.2 and 1.3 are satisfied by $\mathfrak{X}^m = S^m_{\rho\#}$, if $\frac{1}{2} \leq \rho \leq 1$. In case $\frac{1}{2} \leq \rho < 1$, $OP\tilde{\mathfrak{X}}^m$ is a strict subclass of $OPS^m_{\rho,1-\rho}$. It is useful to note that for $\rho = 1$ these two classes coincide.

Proposition 1.4. If $\mathfrak{X}^m = S^m_{1\#}$, then, locally and modulo smoothing operators,

$$(1.52) \qquad OP\tilde{\mathfrak{X}}^m = OPS^m_{1,0}.$$

Proof. We already have $OP\tilde{\mathfrak{X}}^m \subset OPS^m_{1,0}$. For the converse, let $p(x,D) \in OPS^m_{1,0}$. Then, guided by (1.19) and (1.21), we define $\hat{k}(x,\xi)$ by

$$(1.53) \qquad p(x,\xi) = \hat{k}(x,\psi(x,x)\xi)\mathcal{H}_1(x)\tilde{D}(x,x),$$

where ψ, H_1, and D are defined in the proof of proposition 1.1. It is routine to verify that this does indeed define $\hat{k}(x,\xi)$ as a smooth function of x with values in $S^m_{1\#}$. Now define $K_o \in OP\tilde{\mathfrak{X}}^m$ by

(1.54) $$K_o u(x) = \int k(x,xy^{-1})u(y)H_1(y)dy.$$

The proof of proposition 1.1 shows that $K_o \in OPS^m_{1,0}$ and $p(x,D) - K_o = p_2(x,D) \in OPS^{m-1}_{1,0}$. Inductively, we obtain $K_j \in OP\tilde{\mathfrak{X}}^{m-j}$ which asymptotically sum to $K \in OP\tilde{\mathfrak{X}}^m$, and $p(x,D) - K$ is a smoothing operator. This finishes the proof.

We next derive the following simple abstract hypoellipticity theorem from the operator calculus we have developed. In the case of Euclidean space, this is the regularity theorem for elliptic operators.

Proposition 1.5. Suppose $K \in OP\tilde{\mathfrak{X}}^m$ has the form $K = K_1 + K_0$ with $K_0 \in OP\tilde{\mathfrak{X}}^{m-\tau}$ and where $K_1 u(x) = K_1(x)u(x)$. Keep the hypotheses on $\mathfrak{X}^m$ from proposition 1.2. Suppose furthermore that

(1.55) $$E_1(x) = K_1(x)^{-1} \text{ is a smooth function of x with values in } OP\mathfrak{X}^{-m}.$$

Then K is hypoelliptic, with a parametrix in $OP\tilde{\mathfrak{X}}^{-m}$.

Proof. By (1.55), we can define $\mathcal{E}_1 \in OP\tilde{\mathfrak{X}}^{-m}$ by

(1.56) $$\mathcal{E}_1 u(x) = E_1(x)u(x).$$

Then proposition 1.2 implies

(1.57) $$\mathcal{E}_1 K = I - R, \ R \in OP\tilde{\mathfrak{X}}^{-\tau}.$$

Hence

(1.58) $$(I+R+R^2+\cdots+R^N)\mathcal{E}_1 K = I - R^N, \ R^N \in OP\tilde{\mathfrak{X}}^{-N\tau}.$$

Taking N arbitrarily large we deduce K is hypoelliptic, with parametrix

$$\mathcal{E} \sim (1+R+R^2+\cdots)\mathcal{E}_1.$$

It is of interest to consider the behavior of operators under a change of variables, i.e., if $\varphi: G \to G$ is a C^∞ diffeomorphism (not necessarily a group automorphism) with inverse ψ, we want to understand the conjugated operator $\psi^* A\varphi^*$ (where $\varphi^* u(x) = u(\varphi(x))$) in terms of A. A lot of classes of operators to which the previous propositions of this section apply will not be invariant under general diffeomorphisms φ. For example, when $G = R^n$, $\mathfrak{X}^m = S^m_{\rho\#}$, $\frac{1}{2} \le \rho \le 1$, the operator classes $OP\tilde{\mathfrak{X}}^m$ are $OPS^m_{\rho,0}$, but the classes invariant under diffeomorphisms are $OPS^m_{\rho,1-\rho}$, so unless $\rho = 1$, $OP\tilde{\mathfrak{X}}^m$ is not invariant under general diffeomorphisms in this case. Nevertheless, $\mathfrak{X}^m$ in many cases is defined by special properties (e.g., quasi-homogeneity) associated with a certain structure on G (e.g., a contact structure, discussed in Chapter III), and $OP\tilde{\mathfrak{X}}^m$ may be invariant under those diffeomorphisms preserving such a structure. Results along these lines depend strongly on particular cases, but we will outline some general features of the phenomena here, which will apply to special cases, as in Chapter III.

If an operator A is given by

$$Au(x) = \int a(x, xy^{-1}) u(y) \mathcal{H}_1(y)\,dy, \quad x \in G, \tag{1.59}$$

the conjugated operator $B = \psi^* A\varphi^*$ is given by

$$\begin{aligned} Bu(x) &= \int a(\psi(x), \psi(x)y^{-1}) u(\varphi(y)) \mathcal{H}_1(y)\,dy \\ &= \int a(\psi(x), \psi(x)\psi(y)^{-1}) u(y) \tilde{\mathcal{H}}(y)\,dy \\ &= \int b(x, \Psi(x,y,xy^{-1})) u(y) \tilde{\mathcal{H}}(y)\,dy \end{aligned} \tag{1.60}$$

where $b(x,z) = a(\psi(x),z)$ and $\Psi(x,y,xy^{-1}) = \psi(x)\psi(y)^{-1}$. We can arrange that, in exponential coordinates, $\Psi(x,y,z)$ is linear in z. Then

$$\Psi(x,x,z) = D\psi(x)z = \Psi(x)z, \tag{1.61}$$

where $D\psi(x)$ is the derivative of the map ψ, as a linear map on . The next natural thing to do is to make a power series expansion of

$b(x,\Psi(x,y,xy^{-1}))$ in the third argument y, about y = x, in analogy to (1.34). We get

$$(1.62)\qquad b(x,\Psi(x,y,xy^{-1})) \sim \sum_{\substack{\gamma\geq 0\\ |\sigma|\geq|\gamma|}} C_{\sigma\gamma}(xy^{-1})^{\gamma+\sigma} b_{(\gamma)}(x,\Psi(x)(xy^{-1}))$$

where $b_{(\gamma)}(x,z) = D_z^\gamma b(x,z)$. Here, if $w = xy^{-1}$ is given in exponential coordinates by $w = (w_1,\ldots,w_n)$, and if $\gamma = (\gamma_1,\ldots,\gamma_n)$, we set $w^\gamma = w_1^{\gamma_1}\ldots w_n^{\gamma_n}$. Now we would like the summands in (1.62) to be smoothing to high order if $|\gamma|$ is large. Denote the partial Fourier transform of $b(x,w)$ by $\tilde{b}(x,\xi)$. With $w = xy^{-1}$ in (1.62), the partial Fourier transform of a general term in (1.62) with respect to w is

$$(1.63)\qquad C_{\sigma\gamma} D_\xi^{\gamma+\sigma} \xi^\gamma b(x,\Psi(x)^t\xi).$$

If, for example, $\tilde{b}(x,\xi)$ is a smooth function of x with values in $S^m_{1\#}$, then such a term is a smooth function of x with values in $S^{m-|\sigma|}_{1\#} \subset S^{m-|\gamma|}_{1\#}$. Modulo estimating the error term when (1.62) is truncated to a finite sum, this retraces the proof that $OPS^m_{1,0}$ is invariant under coordinate changes. On the other hand, if $\tilde{b}(x,\xi)$ is a smooth function of x with values in $S^m_{\rho\#}$, $\frac{1}{2} < \rho < 1$, then the term in (1.63) belongs to $S^{m+(1-\rho)|\gamma|-\rho|\sigma|}_{\rho,1-\rho} \subset S^{m-(2\rho-1)|\gamma|}_{\rho,1-\rho}$, which is not the same as being a smooth function of x with values in $S^{m-(2\rho-1)|\gamma|}_{\rho\#}$.

We now describe one general class of operators to which many of the results of this section apply. Other operator classes will arise in subsequent chapters. Let G be a simply connected Lie group with a one parameter family $\alpha(t)$ of automorphisms. Denote also by $\alpha(t)$ the automorphisms induced on its Lie algebra $\mathfrak{g}$. Make the assumption that $\alpha(t)$ are dilations, i.e.,

$$(1.64)\qquad \mathfrak{g} = \bigoplus_{b\in\mathcal{B}} \mathfrak{g}_b$$

where $\mathcal{B}$ is a subset of $(0,\infty)$ and

(1.65) $\alpha(t)X = e^{tb}X, \ X \in \mathfrak{g}_b.$

Note that

(1.66) $[\mathfrak{g}_b, \mathfrak{g}_c] \subset \mathfrak{g}_{b+c}.$

Hence, $\mathfrak{g}$ must be nilpotent. Thus, $\exp: \mathfrak{g} \to G$ is a diffeomorphism, taking Lebesque measure on $\mathfrak{g}$ to Haar measure on G.

<u>Definition</u>. $\hat{\mathcal{H}}(G,\alpha,m)$ is the space of distributions u on G such that

(1.67) $u \in C^\infty(G \setminus 0),$

(1.68) $u = u_1 + u_2, \ u_1 \in \mathcal{E}'(G), \ u_2 \in \mathcal{S}(G)$, the Schwartz space,

(1.69) for each $t \in (0,\infty)$, $e^{\sigma t}\alpha(t)^* u - e^{-tm} u \in \mathcal{S}(G)$,

where $e^{\sigma t}$ is the factor by which $\alpha(t)$ expands volumes in G. If $d_j = \dim \mathfrak{g}_{b_j}$, $b_j \in \mathcal{B}$,

(1.70) $\sigma = \Sigma b_j d_j.$

Here, for $v \in \mathcal{S}(G)$, $\alpha(t)^* v(x) = v(\alpha(t)x)$, $x \in G$. Then $\alpha(t)^*$ extends uniquely to a continuous operator on $\mathcal{S}'(G)$. In words, we say u is approximately homogeneous of degree $-(m+\sigma)$. We can characterize the space $\mathcal{H}(G,\alpha,m)$ of Fourier transforms of elements of $\hat{\mathcal{H}}(G,\alpha,m)$ as follows.

<u>Proposition 1.6</u>. u belongs to $\hat{\mathcal{H}}(G,\alpha,m)$ if and only if $\hat{u}(\xi)$ satisfies:

(1.71) $\hat{u} \in C^\infty(\mathfrak{g}')$ (We identify G and $\mathfrak{g}$),

(1.72) $\beta(t)^* \hat{u} - e^{tm}\hat{u} \in \mathcal{S}(\mathfrak{g}')$ for each $t \in (0,\infty)$,

where $\beta(t) = \alpha(t)^t$.

<u>Proof</u>. Since the Fourier transform of $\alpha(t)^* u$ is given by $e^{-\sigma t}\beta(t)^* \hat{u}$, (1.72) and (1.69) are equivalent. Clearly (1.68) implies (1.71). That

(1.71) and (1.72) also give (1.67) follows easily from the following observation.

Proposition 1.7. Suppose $\hat{u} \in C^{\infty}(\mathfrak{g}')$. Then (1.72) is equivalent to the existence of $v \in C^{\infty}(\mathfrak{g}' \setminus 0)$, such that

(1.73) $$v(\beta(t)\xi) = e^{tm}v(\xi) \quad \text{on } \mathfrak{g}' \setminus 0 \text{ for all } t \in (0,\infty),$$

and

(1.74) $$\chi(\xi)[\hat{u}(\xi) - v(\xi)] \in \mathcal{S}(\mathfrak{g}')$$

where $\chi(\xi)$ is smooth, equal to 0 for ξ in a neighborhood of 0, and equal to 1 for ξ outside some compact set.

Proof. Clearly (1.73) and (1.74) imply (1.72). Conversely, suppose

(1.75) $$\hat{u}(\beta(1)\xi) - e^{m}\hat{u}(\xi) = \hat{h}(\xi) \in \mathcal{S}(\mathfrak{g}').$$

Then, replacing ξ by $\beta(j)\xi$,

$$\hat{u}(\beta(j+1)\xi) - e^{m}\hat{u}(\beta(j)\xi) = \hat{h}(\beta(j)\xi).$$

Hence

$$e^{-m(j+1)}\hat{u}(\beta(j+1)\xi) - \hat{u}(\xi) = \sum_{k=0}^{j} e^{-mk}\hat{h}(\beta(k)\xi),$$

so let

(1.76) $$v(\xi) = \sum_{k=0}^{\infty} e^{-mk}\hat{h}(\beta(k)\xi) + \hat{u}(\xi).$$

Since $\hat{h} \in \mathcal{S}(\mathfrak{g}')$, v belongs to $C^{\infty}(\mathfrak{g}' \setminus 0)$, and (1.75) implies

$$e^{-mj}\hat{u}(\beta(j)\xi) \to v(\xi) \text{ as } j \to \infty.$$

Using increments of $\beta(1/N)$ instead of $\beta(1)$ and passing to the limit, we have

(1.77) $$e^{-mt}\hat{u}(\beta(t)\xi) \to v(\xi) \quad \text{as } t \to \infty.$$

It follows that v satisfies (1.73). Furthermore, (1.76) easily yields (1.74), so the proof is complete.

Note that, for $u \in \hat{\mathcal{H}}(G,\alpha,m)$, even though $\hat{u}$ differs by a rapidly decreasing function from a homogeneous function, it is not necessarily the case that u differs by a smooth function from a homogeneous function. Consider the case $G = \mathbb{R}$, $\alpha(t)x = e^t x$. Then

$$\varphi(x) \log|x| \in \hat{\mathcal{H}}(R,e^t,-1), \tag{1.78}$$

where $\varphi \in C_0^\infty(R)$ is equal to 1 near the origin.

Now $OP\mathcal{H}(G,\alpha,m)$ is the set of convolution operators

$$Ku = k*u, \qquad k \in \hat{\mathcal{H}}(G,\alpha,m).$$

We have the following result on compositions of such operators.

<u>Proposition 1.8</u>. If $K_j \in OP\mathcal{H}(G,\alpha,m_j)$, then

$$K_1K_2 \in OP\mathcal{H}(G,\alpha,m_1+m_2). \tag{1.79}$$

<u>Proof</u>. We have $K_j u = k_j*u$, $k_j \in \hat{\mathcal{H}}(G,\alpha,m_j)$. It suffices to show

$$k_1*k_2 \in \hat{\mathcal{H}}(G,\alpha,m_1+m_2). \tag{1.80}$$

This is very simple. Since $\alpha(t)$ is a group of automorphisms of G,

$$\alpha(t)^*(k_1*k_2) = e^{\sigma t}\alpha(t)^*k_1*\alpha(t)^*k_2. \tag{1.81}$$

Now, for each $t \in (0,\infty)$,

$$\alpha(t)^*k_j = e^{-t(m_j+\sigma)}k_j + h_j, \qquad h_j \in \mathcal{S}(G),$$

and since it is elementary to show

$$k_j*h_\ell \text{ and } h_\ell*k_j \in \mathcal{S}(G)$$

under these circumstances, we have, for each $t \in (0,\infty)$,

$$\alpha(t)^*(k_1*k_2) = e^{-t(m_1+m_2+\sigma)}k_1*k_2 \qquad \text{mod } \mathcal{S}(G). \tag{1.82}$$

This completes the proof.

As we have said, an element k of $\hat{\mathcal{H}}(G,\alpha,m)$ need not be equal to a smooth function plus a homogeneous function. However, for some range of m this turns out to be the case. It is useful to note the following simple results on the behavior of elements of $\hat{\mathcal{H}}(G,\alpha,m)$.

Proposition 1.9. If $k \in \hat{\mathcal{H}}(G,\alpha,m)$, then

$$k = k_1 + k_2 \tag{1.83}$$

with $k_2 \in C^\infty(G)$ and k_1 homogeneous, i.e.,

$$\alpha(t)^* k_1 = e^{-t(m+\sigma)} k_1, \tag{1.84}$$

provided

$$m > -\sigma. \tag{1.85}$$

Proof. The point of the hypothesis (1.85) is precisely to guarantee that the homogeneous function $v(\xi)$ of proposition 1.7 be locally integrable in a neighborhood of the origin. This guarantees that $v(\xi)$ defines a homogeneous distribution in $\mathcal{S}'(\mathfrak{G}')$. We can set k_1 equal to the inverse Fourier transform of $v(\xi)$, and (1.83) follows easily.

We take a look at the mapping properties of convolution operators in $OP\mathcal{H}(G,\alpha,0)$. As is well known, such operators are bounded on $L^p(G)$, for $1 < p < \infty$. This follows from the general theories of Coifman and Weiss [C4] and Koranyi and Vagi [K8] which were exploited by [F4] and [R5], and which are given a nice exposition in Goodman [G7]. We give a brief description of the situation. We can define a "homogeneous norm" on G (identified with $\mathfrak{G}$ by the exponential map) by

$$|X| = (\Sigma \|X_i\|^{2/b_i})^{\frac{1}{2}} \tag{1.86}$$

where

$$X = \Sigma X_i, \qquad X_i \in \mathfrak{G}_{b_i},$$

and $\|X_i\|$ is a Euclidean norm on $\mathfrak{G}_{b_i}$. Note that

$$|\alpha(t)X| = e^t|X|. \tag{1.87}$$

If $k \in \hat{\mathcal{H}}(G,\alpha,0)$, then pick k_1 and k_2 as in proposition 1.9, so k_1 is homogeneous of degree $-\sigma$, and smooth outside the origin. We claim k_1 must have mean value zero in each region

$$A_r = \{X \in \mathfrak{g} : r \leq |X| \leq 2r\}. \tag{1.88}$$

In fact, we can pick a unique constant C such that

$$k_1(X) = k_1^{\#}(X) + C|X|^{-\sigma} \tag{1.89}$$

for $X \in \mathfrak{g} \setminus 0$, where $k_1^{\#}$ does have mean value zero on each A_r. Then $k_1^{\#}$ defines a principal value distribution which is homogeneous of degree $-\sigma$ (see [G7]). We can deduce that C in (1.89) is equal to zero if we know that $|X|^{-\sigma}$ cannot be extended to a distribution which is homogeneous of degree $-\sigma$, i.e., whose Fourier transform is homogeneous of degree zero with respect to $\beta(t)$. In fact, not only does $|X|^{-\sigma}$ just fail to be integrable near 0, it also just fails to be integrable near infinity. If a cut-off $\chi(X)$ is used, $\chi \in C^\infty(\mathfrak{g})$, $\chi(X) = 0$ near 0, $\chi(X) = 1$ outside a compact set, $0 \leq \chi(X) \leq 1$ for all X, then $\chi(X)|X|^{-\sigma}$ is smooth, ≥ 0, and not quite integrable at infinity. A simple regularization argument shows its Fourier transform must blow up at the origin. However, such Fourier transform must differ by a smooth function from the Fourier transform of a homogeneous distribution extending $|X|^{-\sigma}$, if such exists, and this is not compatible with homogeneity of order zero for its Fourier transform.

Consequently, each $K \in OP\mathcal{H}(G,\alpha,0)$ is convolution by a k which differs by a smooth function from a k_1 which defines a principal value distribution, homogeneous of degree $-\sigma$, plus a multiple of the delta function concentrated at 0. As shown in [G7], convolution by k_1 is continuous on $L^p(G)$, $1 < p < \infty$. Thus we have:

<u>Proposition 1.10</u>. If $K \in OP\mathcal{H}(G,\alpha,0)$, then $K\colon L^p \to L^p$ for $1 < p < \infty$.

From $OP\mathcal{H}(G,\alpha,m)$ we can pass to the classes of "variable coefficient" operators $OP\tilde{\mathcal{H}}(G,\alpha,m)$, where, as usual, $\mathcal{K} \in OP\tilde{\mathcal{H}}(G,\alpha,m)$ is given by $\mathcal{K}u(x) = K(x)u(x)$ where $K(y)$ is a smooth function of y with values in $OP\mathcal{H}(G,\alpha,m)$. It follows that if $\mathcal{K}_j$ belong to $OP\tilde{\mathcal{H}}(G,\alpha,m_j)$, then $\mathcal{K}_1\mathcal{K}_2$ is in $OP\tilde{\mathcal{H}}(G,\alpha,m_1+m_2)$ and $\mathcal{K}_j^* \in OP\tilde{\mathcal{H}}(G,\alpha,m_j)$. We also have L^p continuity for elements of $OP\tilde{\mathcal{H}}(G,\alpha,0)$:

Proposition 1.11. If $\mathcal{K} \in OP\tilde{\mathcal{H}}(G,\alpha,0)$, then

$$\mathcal{K} : L^p_{comp} \to L^p_{loc} \qquad \text{for } 1 < p < \infty. \tag{1.90}$$

Proof. Using proposition 1.9 we can deduce that, modulo a smoothing operator, $\mathcal{K}$ is given by

$$\mathcal{K}u(x) = k_x * u(x)$$

where k_y is a smooth function of y taking values in $\hat{\mathcal{H}}^0(G,\alpha)$, the space of principal value distributions, homogeneous of degree $-\sigma$, smooth on $\mathfrak{I} \setminus 0$. Working locally, we can suppose $k(y,x) = k_y(x)$ has compact support in y. Now write

$$k(y,x) = \int \ell(\eta,x)e^{iy\cdot\eta}\, d\eta$$

where

$$\ell(\eta,x) = (2\pi)^{-n/2} \int k(y,x)e^{-iy\cdot\eta}dy = \ell_\eta(x).$$

Then ℓ is a rapidly decreasing function of η, taking values in the space $\hat{\mathcal{H}}^0(G,\alpha)$ of principal value distributions. It follows from proposition 1.10 that L_η, defined by $L_\eta u = \ell_\eta * u$, is bounded on $L^p(G)$, with

$$\| L_\eta \|_{\mathcal{L}(L^p)} \leq C_N(p)(1+|\eta|)^{-N}.$$

Since the operator of multiplication by $e^{ix\cdot\eta}$ has operator norm 1 on L^p, the identity

$$\mathcal{K}u(x) = \int e^{ix\cdot\eta} L_\eta u\, d\eta$$

shows that (1.90) holds.

It is desirable to amalgamate some special class of pseudodifferential operators such as $OP\tilde{\mathcal{H}}(G,\alpha,m)$ with the classical pseudodifferential operators OPS^m. We shall describe such an amalgamation in the special case that G is a step 2 nilpotent Lie group, simply connected. In this case, the Fourier integral representation for an operator constructed from convolutions takes the simple form (1.23),(1.24). In particular, if $Ku = k*u$, then

$$Ku = p(x,D)u, \quad p(x,\xi) = \hat{k}(\tilde{L}_x\xi). \tag{1.91}$$

This enables us to establish the following result, following Phong and Stein [P3], who noted it for operators on the Heisenberg group.

<u>Proposition 1.12</u>. Suppose G is a step 2 nilpotent Lie group. If $Ku = k*u$ defines a convolution operator in $OP\mathcal{H}(G,\alpha,m)$ and $Lu = \ell*u$ defines a convolution operator in OPS^{μ}, then

$$LK\,u = Pu = p*u \tag{1.92}$$

where $\hat{p}(\xi)$ has an asymptotic expansion of the form

$$\hat{p}(\xi) \sim \hat{k}(\xi)\hat{\ell}(\xi) + \sum_{j\geq 1} a_j(\xi)b_j(\xi). \tag{1.93}$$

Here,

$$a_j(\xi) \in \mathcal{H}(G,\alpha,m_j), \quad b_j(\xi) \in S^{\mu_j}_{\#}(\mathfrak{g}') \tag{1.94}$$

with

$$m_j \leq m, \ \mu_j \leq \mu, \ m_j+\mu_j < m+\mu, \text{ and } m_j+\mu_j \to -\infty \text{ as } j \to \infty. \tag{1.95}$$

Also

$$KL\,u = P_1u = p_1*u \tag{1.96}$$

where $\hat{p}_1(\xi)$ has an expansion similar to (1.93), with the same leading term $\hat{k}(\xi)\hat{\ell}(\xi)$.

Proof. By (1.91), convolution operators are characterized by

$$Au = A(x,D)u = \int A(x,\xi)\hat{u}(\xi)e^{ix\cdot\xi}d\xi$$

with

$$A(x,\xi) = \hat{a}(\tilde{L}_x\xi).$$

If we put K in this form, we have

$$(1.97) \qquad Ku = K(x,D)u = \int K(x,\xi)\hat{u}(\xi)e^{ix\cdot\xi}d\xi$$

with

$$K(x,\xi) = \hat{k}(\tilde{L}_x\xi).$$

Now if $\hat{k} \in \mathcal{H}(G,\alpha,m)$, it follows that

$$(1.98) \qquad \hat{k}(\xi) \in S^M_{\rho\#}$$

with $\rho = \rho(b_1,\ldots)$ and

$$(1.99) \qquad M = m \max_j b_j, \text{ for } m \geq 0, \ M = m \min_j b_j \quad \text{for } m < 0.$$

The standard pseudodifferential operator calculus allows us to apply

$$(1.100) \qquad L = L(x,D) \in OPS^{\mu} \qquad (L(x,\xi) = \hat{\ell}(\tilde{L}_x\xi))$$

to (1.97), to get

$$(1.101) \qquad LK = P(x,D) \in OPS^{M+\mu}_{\rho,1-\rho}$$

with

$$(1.102) \qquad P(x,\xi) \sim \sum_{\alpha\geq 0} (1/\alpha!)L^{(\alpha)}(x,\xi)D^{\alpha}_x K(x,\xi).$$

Since LK is a convolution operator, $P(x,\xi) = \hat{p}(\tilde{L}_x\xi) = P(0,\tilde{L}_x\xi)$, so evaluating (1.102) at $x = 0$ gives (1.93). The analysis of KL is the same.

Proposition 1.12 motivates considering the following classes of symbols.

Definition. $\mathcal{H}^{m,\mu}_{\alpha,\delta}$ is the class of symbols $p(\xi)$ with asymptotic expansion

(1.103) $$p(\xi) \sim \sum_{j\geq 0} a_j(\xi) b_j(\xi)$$

where

(1.104) $$a_j(\xi) \in \mathcal{H}(G,\alpha,m_j),\ b_j(\xi) \in \Sigma^{\mu_j}(\mathfrak{g}')$$

and

(1.105) $$m_j \leq m,\ \mu_j \leq \mu,\ m_j+\mu_j \to -\infty \text{ as } j \to \infty.$$

Here $\Sigma^\mu = S^\mu_{\#}$ denotes the subspace of $S^\mu_{1\#}$ consisting of $b(\xi) \in S^\mu_{1\#}$, with asymptotic expansion $b(\xi) \sim \Sigma b_j(\xi)$, b_j homogeneous of degree $\mu-j$ in ξ, with respect to the homogeneous dilations

(1.106) $$\delta(t)\xi = e^t\xi.$$

From now on, $OP\Sigma^\mu$ will denote the set of convolution operators $Lu = \ell * u$ which belong to OPS^μ. Proposition 1.12 shows that, for G nilpotent of step 2,

(1.107) $$K \in OP\mathcal{H}(G,\alpha,m),\ L \in OP\Sigma^\mu \Rightarrow KL \text{ and } LK \in OP\mathcal{H}^{m,\mu}_{\alpha,\delta}.$$

We have the following converse result.

Proposition 1.13. If G is a step 2 nilpotent Lie group and $P \in OP\mathcal{H}^{m,\mu}_{\alpha,\delta}$, then P has an asymptotic expansion

(1.108) $$P \approx \sum_{j\geq 0} K_j L_j,$$

in the sense that the difference of P and a sufficiently large finite partial sum is smoothing to arbitrary order, where

(1.109) $$K_j \in OP\mathcal{H}(G,\alpha,m_j),\ L_j \in OP\Sigma^{\mu_j},$$

with

(1.110) $$m_j \leq m,\ \mu_j \leq \mu, \text{ and } m_j+\mu_j \to -\infty.$$

Proof. We can reduce our problem to considering

(1.111) $$\hat{p}(\xi) = \hat{k}_0(\xi)\hat{\ell}_0(\xi),\ \ \hat{k}_0 \in \mathcal{H}(G,\alpha,m),\ \hat{\ell}_0 \in \Sigma^\mu.$$

Then $K_0 u = k_0 * u$ defines an operator $K_0 \in OPH(G,\alpha,m)$, and $L_0 u = \ell_0 * u$ defines $L_0 \in OP\Sigma^{\mu}$. If we apply proposition 1.12 to $K_0 L_0$ we see that $K_0 L_0 \in OPH^{m,\mu}_{\alpha,\delta}$ and the principal term in the expansion is given by (1.111). Applying the same reasoning to the remainder terms and iterating repeatedly, we get (1.108).

This result enables us to obtain the following L^p boundedness.

Proposition 1.14. If G is a step 2 nilpotent Lie group and $P \in OPH^{0,0}_{\alpha,\delta}$, then

$$P: L^p \to L^p \text{ for } 1 < p < \infty.$$

Proof. By (1.108), this follows from proposition 1.10, which gives

$$K_j: L^p \to L^p,\ 1 < p < \infty,\ \text{if } K_j \in OPH(G,\alpha,m_j),\ m_j \leq 0$$

(the result for $m_j < 0$ being elementary), together with the well known result

$$L_j: L^p \to L^p,\ 1 < p < \infty \text{ if } L_j \in OPS^{\mu_j},\ \mu_j \leq 0.$$

This result was proved by Phong and Stein [P3], in the context of convolution operators on the Heisenberg group. They also treated the harder problem of weak type (1,1) properties, which are not necessarily preserved by taking compositions, as are L^p continuity properties.

The operator clsses $OPH^{m,\mu}_{\alpha,\delta}$ have the following properties for compositions and adjoints:

Proposition 1.15. If G is step 2 and $P_j \in OPH^{m_j,\mu_j}_{\alpha,\delta}$, then

(1.112) $$P_1 P_2 \in OPH^{m_1+m_2,\mu_1+\mu_2}_{\alpha,\delta}$$

and

(1.113) $$P_j^* \in OPH^{m_j,\mu_j}_{\alpha,\delta}.$$

Proof. Apply (1.108) to P_j. Then $P_1 P_2$ is a sum of terms

(1.114) $$K_kL_kK'_kL'_k, \quad K_k \in OP\mathcal{H}(G,\alpha,m_k),\ K'_k \in OP\mathcal{H}(G,\alpha,m'_k), \text{ etc.}$$

where $m_k \leq m_1$, $m'_k \leq m_2$, etc. Now (1.107) implies $L_kK'_k \in OP\mathcal{H}^{m_k,\mu'_k}_{\alpha,\delta}$, and hence, by proposition 1.13,

(1.115) $$L_kK'_k \sim \sum_{\nu\geq 0} K'_{k\nu}L_{k\nu}.$$

If we substitute (1.115) into (1.114), we see that (1.112) holds for P_1P_2. The same sort of argument yields (1.113).

We now pass from the classes of convolution operators $OP\mathcal{H}^{m,\mu}_{\alpha,\delta}$ to their "variable coefficient" versions $OP\tilde{\mathcal{H}}^{m,\mu}_{\alpha,\delta}$. Since $\mathcal{H}^{m,\mu}_{\alpha,\delta}$ does not have a Frechet space structure, we make the special definition that an operator $\mathcal{K}u(x) = K(x)u(x)$ belongs to $OP\tilde{\mathcal{H}}^{m,\mu}_{\alpha,\delta}$ if

(1.116) $$\hat{k}(y,\xi) \sim \sum_{j\geq 0} a_j(y,\xi)b_j(y,\xi)$$

where $a_j(y,\xi)$ is a smooth function of y with values in $\mathcal{H}(G,\alpha,m_j)$, $b_j(y,\xi)$ is a smooth function of y with values in $\Sigma^{\mu_j} = S^{\mu_j}_{\#}$, and, as in (1.105),

(1.117) $$m_j \leq m,\ \mu_j \leq \mu,\ m_j+\mu_j \to -\infty \text{ as } j \to \infty.$$

The meaning of this is that, for any K, if N is large enough,

(1.118) $$\hat{k}(y,\xi) - \sum_{j\leq N} a_j(y,\xi)b_j(y,\xi) \in S^{-K}_{\rho\#}, \text{ smoothly in } y,$$

where ρ is some number in $(0,1]$. The basic operator calculi developed so far apply. We have the following consequences of proposition 1.12 and proposition 1.15.

<u>Proposition 1.16</u>. If G is step 2 and $\mathcal{P}_j \in OP\tilde{\mathcal{H}}^{m_j,\mu_j}_{\alpha,\delta}$, then

(1.119) $$\mathcal{P}_1\mathcal{P}_2 \in OP\tilde{\mathcal{H}}^{m_1+m_2,\mu_1+\mu_2}_{\alpha,\delta},$$

and

(1.120) $$\mathcal{P}^*_j \in OP\tilde{\mathcal{H}}^{m_j,\mu_j}_{\alpha,\delta}.$$

We are also able to obtain the following, in parallel with proposition 1.13:

Proposition 1.17. If G is step 2 and $\mathcal{P} \in OP\tilde{\mathcal{H}}^{m,\mu}_{\alpha,\delta}$, then $\mathcal{P}$ has an asymptotic expansion

(1.121) $$\mathcal{P} \sim \sum_{j\geq 0} \mathcal{K}_j \mathcal{L}_j$$

where

(1.122) $$\mathcal{K}_j \in OP\tilde{\mathcal{H}}(G,\alpha,m_j), \quad \mathcal{L}_j \in OPS^{\mu_j}$$

with

(1.123) $$m_j \leq m, \ \mu_j \leq \mu, \ m_j+\mu_j \to -\infty.$$

Proof. This goes like the proof of proposition 1.13. We can reduce our problem to considering

(1.124) $$\hat{p}_y(\xi) = \hat{k}_y(\xi)\hat{\ell}_y(\xi), \quad \hat{k}_y(\xi) \in \mathcal{H}(G,\alpha,m), \quad \hat{\ell}_y(\xi) \in \Sigma^{\mu}.$$

Then $\mathcal{K}_o u = k_x * u(x)$ defines an operator $\mathcal{K}_o \in OP\tilde{\mathcal{H}}(G,\alpha,m)$ and $\mathcal{L}_o u = \ell_x * u(x)$ defines $\mathcal{L}_o \in OPS^{\mu}$. If we apply proposition 1.16 to $\mathcal{K}_o\mathcal{L}_o$ we see that $\mathcal{K}_o\mathcal{L}_o$ belongs to $OP\tilde{\mathcal{H}}^{m,\mu}_{\alpha,\delta}$ and the principal term in the expansion is given by (1.124). Applying the same reasoning to the remainder term and iterating repeatedly, we get (1.121).

As in proposition 1.14, we deduce the following L^p boundedness.

Proposition 1.18. If G is step 2 and $\mathcal{P} \in OP\tilde{\mathcal{H}}^{0,0}_{\alpha,\delta}$, then

(1.125) $$\mathcal{P}: L^p_{comp} \to L^p_{loc} \quad \text{for } 1 < p < \infty.$$

Proof. Using the decomposition (1.121), it suffices to invoke proposition 1.11 and the well known L^p boundedness of $OPS^0_{1,0}$.

2. Symbols.

Section 1 of this chapter has developed what one would call an "operator calculus." It is one of our main goals in this paper to elevate this to a "symbol calculus." The symbol of an operator in $OP\tilde{\mathfrak{X}}$ is described in terms of the irreducible unitary representations of G.

If $\hat{G}$ denotes the set of (equivalence classes of) such representations, then the symbol of a convolution operator $Ku = k*u$ will be defined as

$$\sigma_K(x,\pi) = \pi(k), \quad x \in \hat{G}, \quad \pi \in \hat{G}, \tag{2.1}$$

where

$$\pi(k) = \int_G k(y)\pi(y)\mathcal{H}(y)dy. \tag{2.2}$$

At least, if $k \in L^1(G)$, (2.2) is well defined and is bounded operator on the representation space H_π of π. We briefly set up machinery to define $\pi(k)$ for more singular $k \in \mathcal{E}'(G)$.

A vector $v \in H_\pi$ is called a smooth vector (we write $v \in H_\pi^\infty$) if the function $\varphi_v(x) = \pi(x)v$ is C^∞ from G to H_π. Following Rockland [R2], we call v weakly smooth (we write $v \in H_\pi^{\infty w}$) if the function $\varphi_{v,w}(x) = (\pi(x)v,w)$ is C^∞ on G for each $w \in H_\pi$. As is well known, H_π^∞ contains the Gårding space $\{\pi(\psi)\colon \psi \in C_o^\infty(G)\}$ and consequently is dense in H_π. Clearly $H_\pi^\infty \subset H_\pi^{\infty w}$. As a matter of fact, these two spaces coincide:

$$H_\pi^\infty = H_\pi^{\infty w}. \tag{2.3}$$

To see this, let $\Phi(x)$ be a function from an open set U in G (which we identify with a ball in $\mathbb{R}^n$) to H_π which is weakly C^1. Hence, for $w \in H_\pi$, $x,y \in U$,

$$(\Phi(x+y),w) = (\Phi(x),w) + \Sigma y_j D_j(\Phi(x),w) + R_1(x,y,w) \tag{2.4}$$

where $|y|^{-1}R_1(x,y,w) \to 0$ as $|y| \to 0$ in $\mathbb{R}^n$. The uniform boundedness theorem implies

$$|y|^{-1} \; \|\Phi(x+y) - \Phi(x)\| \le C_1.$$

This implies $\Phi(x)$ is strongly Lipschitz (in particular strongly continuous). This gives $|D_j(\Phi(x),w)| \le C_1\|w\|$, and hence there are $\Psi_j(x) \in H_\pi$ such that $D_j(\Phi(x),w) = (\Psi_j(x),w)$; $\|\Psi_j(x)\| \le C_1$. Now if Φ is weakly C^2, then each Ψ_j is weakly C^1 and by the argument just given Ψ_j is continuous. Furthermore, if we write

$$(\Phi(x+y),w) = (\Phi(x),w) + \Sigma y_j(\Psi_j(x),w) + D^2(\Phi(x),w)\cdot y\cdot y + R_2(x,y,w)$$

where $|y|^{-2} R_2(x,y,w) \to 0$ as $|y| \to 0$ and apply the uniform boundedness theorem, we get

$$|y|^{-2} \|\Phi(x+y) - \Phi(x) = \Sigma y_j\Psi_j(x)\| \le C_2.$$

This implies Φ is strongly differentiable, with derivative $(\Psi_1(x),\dots,\Psi_n(x))$. Thus, if Φ is weakly C^2 it is strongly C^1. It follows easily that if Φ is weakly C^k it is strongly C^{k-1}, and this gives (2.3).

Now, given $k \in \mathcal{E}'(G)$, we define

(2.5) $$\pi(k)\colon H_\pi^\infty \to H_\pi$$

by

(2.6) $$(\pi(k)v,w) = \langle \varphi_{v,w}, k\rangle.$$

It is routine to check this is well defined. Also, for $v \in H_\pi^\infty$, $w \in H_\pi$,

$$(\pi(x)\pi(k)v,w) = \langle \varphi_{v,w}, \rho_x k\rangle$$

where $\rho_x k(y) = k(x^{-1}y)$ for $k \in C_o^\infty$, extending to $k \in \mathcal{E}'$, and this makes it clear that $\pi(k)v \in H_\pi^{\infty w}$. In view of (2.3), we have

(2.7) $$\pi(k)\colon H_\pi^\infty \to H_\pi^\infty \quad \text{for } k \in \mathcal{E}'(G).$$

It is routine to verify that, for $k_1,k_2 \in \mathcal{E}'(G)$, $k_1*k_2 \in \mathcal{E}'(G)$, and

(2.8) $$\pi(k_1*k_2) = \pi(k_1)\pi(k_2),$$

and, if $\check{k}(x) = \overline{k(x^{-1})}$ for $k \in C_o^\infty$, then $k \mapsto \check{k}$ extends to $k \in \mathcal{E}'$, and

(2.9) $$\pi(\check{k}) = \pi(k)^*.$$

With these preliminaries out of the way, we define the symbol of an operator

(2.10) $$\mathcal{K}u(x) = \int k(x,xy^{-1})u(y)\mathcal{H}_1(y)dy$$

for $k(x,\cdot)$ a smooth function of x with values in $\mathcal{E}'(G)$, by

(2.11) $$\sigma_{\mathcal{K}}(x,\pi) = \pi(k_x)$$

where $k_x(y) = k(x,y)$. Thus, loosely speaking,

(2.12) $$\sigma_{\mathcal{K}}(x,\pi) = \int_G k(x,y)\pi(y)\mathcal{H}(y)dy.$$

By (2.8), we have the following formula for the symbol of the term $\mathcal{C}_\gamma$,

(2.13) $$\mathcal{C}_\gamma u(x) = A^{[\gamma]}(x)B_{[\gamma]}(x)u(x)$$

in the expansion (1.41) for a composite map $\mathcal{AB}$:

(2.14) $$\sigma_{\mathcal{C}_\gamma}(x,\pi) = \pi(A^{[\gamma]}(x))\pi(B_{[\gamma]}(x)).$$

Thus we can rewrite (1.41) as

(2.15) $$\sigma_{\mathcal{AB}}(x,\pi) \sim \sum_{\gamma\geq 0} \pi(A^{[\gamma]}(x))\pi(B_{[\gamma]}(x)).$$

Recall that $\mathcal{AB}-\mathcal{C} \in OP\tilde{\mathcal{X}}^{m+\mu-\tau}$, and we have

(2.16) $$\sigma_{\mathcal{C}}(x,\pi) = \sigma_{\mathcal{A}}(x,\pi)\sigma_{\mathcal{B}}(x,\pi).$$

Similarly, on the symbol level, we can write the asymptotic expansion (1.51) for the adjoint $\mathcal{K}^*$ of $\mathcal{K} \in OP\tilde{\mathcal{X}}^m$ as

(2.17) $$\sigma_{\mathcal{K}^*}(x,\pi) \sim \sum_{\gamma\geq 0} \pi(K_{\{\gamma\}}(x)).$$

In this case, $\mathcal{K}^* - \mathcal{K}^\# \in OP\tilde{\mathcal{X}}^{m-\tau}$ with

(2.18) $$\sigma_{\mathcal{K}^\#}(x,\pi) = \sigma_{\mathcal{K}}(x,\pi)^*.$$

The problem that remains in producing genuine symbol calculi is to specify classes $\mathfrak{X}^m$, satisfying (1.25)-(1.27), (1.42)-(1.43), and (1.46), and to describe such classes in terms of their symbols. Important classes of such symbols will arise in the following chapters.

We make some further remarks about the continuity of operators on L^2. If G is a unimodular type I group, there is a Plancherel measure $d\mu$ on $\hat{G}$ such that

$$(2.19) \qquad \|u\|^2_{L^2(G)} = \int_{\hat{G}} \|\pi(u)\|^2_{HS}\, d\mu(\pi).$$

Consequently, a convolution operator $Ku = k*u$ is continuous on $L^2(G)$ if and only if the operator norms of $\pi(k)$ are uniformly bounded, as π ranges over $\hat{G}$. As for operators of the form (1.10), (1.11), we have the following result.

Proposition 2.1. Let $\mathfrak{X}$ be a Frechet space such that for each convolution operator $Ku = k*u$, $K \in OP\mathfrak{X}$, the operator norms $\|\pi(k)\|$ are uniformly bounded as π runs over $\hat{G}$. Then each $K \in OP\tilde{\mathfrak{X}}$ is continuous on L^2, provided G is a unimodular type I group.

Proof. We can regard

$$(2.20) \qquad K_o u(x) = K(y)u(x) = v(y,x)$$

as a convolution operator on G from scalar functions to functions taking values in the Sobolev space $H^K(G_y)$, with K picked arbitrarily large. Since the operator norms of $\pi(D_y^\alpha k(y,\cdot))$ are uniformly bounded as π runs over $\hat{G}$, and y runs over a neighborhood of $e \in G$, for each α, it follows that

$$(2.21) \qquad K_o : L^2(G) \to L^2(G, H^K).$$

If K is picked large enough to apply the Sobolev imbedding theorem, we deduce that for $u \in L^2$, $Ku(x) = v(x,x)$ also belongs to L^2, which completes the proof.

Let us note that, if G is a type I unimodular group, the Plancherel formula (2.19) polarizes to give

$$(u,v) = \int \mathrm{tr}(\pi(v)^* \pi(u))\, d\mu(\pi). \tag{2.22}$$

If we let v be an approximate identity and pass to the limit, and if we have control over the trace norm of $\pi(u)$, we obtain the inversion formula

$$u(x) = \int_{\hat{G}} \mathrm{tr}(\pi(x)^* \pi(u))\, d\mu(\pi), \quad x \in G. \tag{2.23}$$

Now, since $(k*u) = \pi(k)\pi(u)$, we get the following formula for the operator $\mathbb{K}$ given by (2.10)-(2.12) (granted that, for $u \in C_0^\infty(G)$, $\|\pi(u)\|_{tr}$ is under control):

$$\mathbb{K}u(x) = \int_{\hat{G}} \mathrm{tr}(\pi(x)^* \sigma_{\mathbb{K}}(x,\pi)\pi(u))\, d\mu(\pi). \tag{2.24}$$

Note the direct parallel with the formula for a pseudodifferential operator on $\mathbb{R}^n$:

$$p(x,D)u(x) = \int_{R^n} e^{ix\cdot\xi} p(x,\xi)\hat{u}(\xi)\, d\xi. \tag{2.25}$$

Actually, however neat this parallel is, we will not make much direct use of the formula (2.24).

Chapter II. Harmonic Analysis on the Heisenberg Group

The Heisenberg group $\mathbb{H}^n$ has the simplest representation theory of all noncommutative Lie groups. We develop several classes of right invariant pseudodifferential operators on $\mathbb{H}^n$. A point worth emphasizing is that we give a unified treatment of convolution kernels with the type of homogeneity related to certain automorphisms of $\mathbb{H}^n$ and also of the classical pseudodifferential operators, with the usual Euclidean homogeneity. Using harmonic analysis on $\mathbb{H}^n$ to study this last class is not as direct as studying OPS^m via Fourier analysis on $\mathbb{R}^n$, but once one puts forth the effort to do it this way, one is rewarded with natural amalgamations of classes of operators with different types of homogeneity. This facilitates doing microlocal analysis on the Heisenberg group.

1. Convolution operators on the Heisenberg group

The purpose of this section is to achieve a basic understanding of the symbol $\pi(k)$ of a convolution operator $Ku = k*u$ where k is a compactly supported function (or distribution) on the Heisenberg group. The material in the first half of this section is well known; we collect it for use in the next section. We begin with a brief description of the Heisenberg group $\mathbb{H}^n$ and its irreducible unitary representations. For a more complete discussion, see the first chapter of [T5].

As a C^∞ manifold, $\mathbb{H}^n$ is $\mathbb{R}^{2n+1}$. Let us denote a point in $\mathbb{H}^n$ by (t,q,p); $t \in \mathbb{R}$, q and $p \in \mathbb{R}^n$. The group law is given by

(1.1) $$(t_1,q_1,p_1)\cdot(t_2,q_2,p_2) = (t_1+t_2+\tfrac{1}{2}p_1\cdot q_2-\tfrac{1}{2}p_2\cdot q_1,\ q_1+q_2,\ p_1+p_2).$$

The Lie algebra $\mathfrak{h}^n$ of $\mathbb{H}^n$ is spanned by the right invariant vector fields

(1.2) $$T = \partial/\partial t,\ L_j = \partial/\partial q_j - \tfrac{1}{2}p_j\partial/\partial t,\ M_j = \partial/\partial p_j + \tfrac{1}{2}q_j\partial/\partial t,\ 1\le j\le n.$$

Note that

$$(1.3)\qquad [L_j,M_j] = -[M_j,L_j] = T,$$

all other commutators being zero.

For $\lambda \in (0,\infty)$, irreducible unitary representations of $\mathbb{H}^n$ on $L^2(\mathbb{R}^n)$ are given by

$$(1.4)\qquad \pi_{\pm\lambda}(t,q,p) = e^{i(\pm\lambda tI \pm \lambda^{\frac{1}{2}} q\cdot X + \lambda^{\frac{1}{2}} p\cdot D)}$$

where $q\cdot X$ is a multiplication operator defined by

$$(1.5)\qquad (q\cdot X)u(x) = \Sigma q_j x_j u(x),$$

and $p\cdot D$ is a differential operator given by

$$(1.6)\qquad (p\cdot D)u(x) = (1/i)\Sigma p_j \partial u/\partial x_j .$$

An alternative formula, equivalent to (1.4), is

$$(1.7)\qquad \pi_{\pm\lambda}(t,q,p)u(x) = e^{i(\pm\lambda t \pm \lambda^{\frac{1}{2}} q\cdot x \pm \lambda q\cdot p/2)} u(x+\lambda^{\frac{1}{2}} p).$$

There are also one dimensional representations $\pi_{(y,\eta)}$, $(y,\eta) \in \mathbb{R}^{2n}$, given by

$$(1.8)\qquad \pi_{(y,\eta)}(t,q,p)v = e^{i(y\cdot q + \eta\cdot p)} v, \qquad v \in \mathbb{C}.$$

It is the content of the Stone-von Neumann theorem that any irreducible unitary representation of $\mathbb{H}^n$ is unitarily equivalent to one of those just described.

The Plancherel identity on $\mathbb{H}^n$ is

$$(1.9)\qquad \int_{\mathbb{H}^n} |u(x)|^2 dx = c_n \int_{\mathbb{R}\setminus 0} \|\pi_\lambda(u)\|_{HS}^2 \, |\lambda|^n d\lambda,$$

where $\|T\|_{HS}^2 = \mathrm{tr}(T^* T)$ is the squared Hilbert-Schmidt norm of T. (Haar measure on $\mathbb{H}^n$ coincides with Lebesgue measure on $\mathbb{R}^{2n+1}$.) A proof of (1.9) will be given below. In particular, the set of one-dimensional representations (1.8) has Plancherel measure zero.

Given a compactly supported function (or distribution) k on $\mathbb{H}^n$, we want to understand $\pi_{\pm\lambda}(k)$ and $\pi_{(y,\eta)}(k)$. Indeed, (1.4) gives

$$\text{(1.10)} \qquad \pi_{\pm\lambda}(k) = \int k(t,q,p)e^{i(\pm\lambda tI\pm\lambda^{\frac{1}{2}}q\cdot X+\lambda^{\frac{1}{2}}p\cdot D)}dt\,dq\,dp$$

$$= \hat{k}(\pm\lambda,\pm\lambda^{\frac{1}{2}}X,\lambda^{\frac{1}{2}}D),$$

where $\hat{k}(\tau,y,\eta)$ denotes the Euclidean space Fourier transform

$$\text{(1.11)} \qquad \hat{k}(\tau,y,\eta) = \int k(t,q,p)e^{i(t\tau+q\cdot y+\eta\cdot p)}dt\,dq\,dp$$

and the operator a(X,D) is defined by the Weyl functional calculus:

$$\text{(1.12)} \qquad a(X,D) = \int \tilde{a}(q,p)e^{i(q\cdot X+p\cdot D)}dq\,dp$$

($\tilde{a}(q,p)$ is the inverse Fourier transform of a). Such operators have been studied by several people, including Grossman, Loupias and Stein [G11], Voros [V3], Hörmander [H10], and Howe [H11]; see also Nelson [N4], and Anderson [A1]. A few manipulations of integrals give the following formula for a(X,D):

$$\text{(1.13)} \qquad a(X,D)u(x) = (2\pi)^{-n}\iint e^{i(x-y)\cdot\xi}a(\tfrac{1}{2}(x+y),\xi)u(y)\,dy\,d\xi.$$

To restate (1.10), we have

$$\text{(1.14)} \qquad \pi_{\pm\lambda}(k) = \sigma_K(\pm\lambda)(X,D)$$

where

$$\text{(1.15)} \qquad \sigma_K(\pm\lambda)(x,\xi) = \hat{k}(\pm\lambda,\pm\lambda^{\frac{1}{2}}x,\lambda^{\frac{1}{2}}\xi),$$

or equivalently

$$\text{(1.16)} \qquad \hat{k}(\pm\tau,y,\eta) = \sigma_K(\pm\tau)(\pm\tau^{-\frac{1}{2}}y,\tau^{-\frac{1}{2}}\eta), \qquad \tau > 0.$$

The behavior of $\pi_{(y,\eta)}(k)$ is given simply by

$$\text{(1.17)} \qquad \pi_{(y,\eta)}(k) = \int k(t,q,p)e^{i(y\cdot q+\eta\cdot p)}dt\,dq\,dp$$

$$= \hat{k}(0,y,\eta).$$

Regarding the associated representation of the Lie algebra $\mathfrak{h}^n$ of $\mathbb{H}^n$, we have

$$(1.18)\qquad \pi_{\pm\lambda}(T) = \pm i\lambda,\ \pi_{\pm\lambda}(L_j) = \pm i\lambda^{\frac{1}{2}}x_j,\ \pi_{\pm\lambda}(M_j) = \lambda^{\frac{1}{2}}\partial/\partial x_j,$$

and

$$(1.19)\qquad \pi_{(y,\eta)}(T) = 0,\ \pi_{(y,\eta)}(L_j) = iy_j,\ \pi_{(y,\eta)}(M_j) = i\eta_j.$$

We now show how formulas (1.10)-(1.13) give a proof of the Plancherel formula (1.9). Note that the squared Hilbert-Schmidt norm of an operator $a(X,D)u(x) = \int A(x,y)u(y)dy$ is $\iint |A(x,y)|^2 dx\,dy$. Thus (1.10) and (1.13) imply

$$(1.20)\qquad c_n \|\pi_{\pm\lambda}(k)\|^2_{HS} = \int_{R^{2n}} |\hat{k}(\pm\lambda, \pm\lambda^{\frac{1}{2}}x, \lambda^{\frac{1}{2}}\xi)|^2\,dx\,d\xi$$

$$= |\lambda|^{-n} \int_{R^{2n}} |\hat{k}(\pm\lambda,y,\eta)|^2\,dy\,d\eta.$$

so

$$(1.21)\qquad c_n \int_{-\infty}^{\infty} \|\pi_\lambda(k)\|^2_{HS}\,|\lambda|^n\,d\lambda = \int_{R^{2n+1}} |\hat{k}(\lambda,y,\eta)|^2\,d\lambda\,dy\,d\eta.$$

Now (1.9) follows from this, together with the ordinary Euclidean space Plancherel theorem

$$(1.22)\qquad \int_{\mathbb{H}^{n+1}} |k(z)|^2 dz = \int_{R^{2n+1}} |\hat{k}(\lambda,y,\eta)|^2\,d\lambda\,dy\,d\eta.$$

This completes the proof of (1.9). Note that polarization of (1.9) gives

$$(1.23)\qquad \int_{\mathbb{H}^n} f(z)\overline{g(z)}\,dz = c_n \int_{-\infty}^{\infty} \operatorname{tr}(\pi_\lambda(g)^* \pi_\lambda(f))\,|\lambda|^n\,d\lambda.$$

If we replace g by a sequence in $C_0^\infty(\mathbb{H}^n)$ tending to the delta function and pass to the limit, we get the inversion formula

$$(1.24)\qquad f(z) = c_n \int_{-\infty}^{\infty} \operatorname{tr}(\pi_\lambda(z)^* \pi_\lambda(f))\,|\lambda|^n\,d\lambda.$$

We can use the Plancherel formula to estimate the operator norm of a convolution operator $Ku = k*u$. In fact, (1.9) implies

(1.25) $$\| K \|_{\mathcal{L}(L^2)} \leq \sup_{\lambda} \|\sigma_K(\pm\lambda)(X,D)\|,$$

where the latter norm is the operator norm on $L^2(\mathbb{R}^n)$. We can estimate this operator norm via the following special case of the Calderon-Vaillancourt theorem. Suppose $a(x,\xi)$ satisfies the estimates

(1.26) $$|D^{\alpha}_{x,\xi}a(x,\xi)| \leq A, \ |\alpha| \leq K(n),$$

where $K(n)$ is sufficiently large. Then

(1.27) $$\| a(X,D) \| \leq C(n)A.$$

A proof of this is given in Appendix A at the end of this paper. In light of (1.25) and (1.15), this implies the following estimate on $\|K\|$:

(1.28) $$\| K \|_{\mathcal{L}(L^2)} \leq C(n) \sup_{\pm\lambda,y,\eta} \sup_{|\alpha|\leq K(n)} \lambda^{|\alpha|/2}|D^{\alpha}_{y,\eta}\hat{k}(\pm\lambda,y,\eta)|.$$

In particular, $K: L^2(\mathbb{H}^n) \to L^2(\mathbb{H}^n)$ if $\hat{k} \in S^0_{\frac{1}{2}\#}$, where recall from Chapter I that $\hat{k} \in S^m_{\frac{1}{2}\#}$ means

(1.29) $$|D^{\alpha}_{\tau,y,\eta}\hat{k}(\tau,y,\eta)| \leq C_{\alpha}(1+|\tau|+|y|+|\eta|)^{m-\frac{1}{2}|\alpha|}.$$

Of course, proposition 1.1 from Chapter I implies $K \in OPS^0_{\frac{1}{2},\frac{1}{2}}$ in this case, so the L^2 boundedness of K when $\hat{k} \in S^0_{\frac{1}{2}\#}$ also follows directly from the Calderon-Vaillancourt theorem for $OPS^0_{\frac{1}{2},\frac{1}{2}}$, as already noted in Chapter I.

One significant structure that accompanies the Heisenberg group is the family of dilations

(1.30) $$\delta_{\pm\lambda}(t,q,p) = (\pm\lambda t, \pm\lambda^{\frac{1}{2}}q, \lambda^{\frac{1}{2}}p), \quad \lambda > 0.$$

These are all automorphisms of $\mathbb{H}^n$. Note that

(1.31) $$\pi_{\pm\lambda}(w) = \pi_1(\delta_{\pm\lambda}w), \ w \in \mathbb{H}^n.$$

If we let δ_λ^* act on $C_o^\infty(\mathbb{H}^n)$ (or on $\mathcal{E}'(\mathbb{H}^n)$ or other spaces) by

$$\delta_\lambda^* u(w) = u(\delta_\lambda w) \tag{1.32}$$

then

$$\pi_\lambda(\delta_\tau^* u) = |\tau|^{-n-1}\pi_{\lambda/\tau}(u). \tag{1.33}$$

We make a few remarks on the space $\mathcal{S}(\mathbb{H}^n)$ of functions which, together with all their derivatives, are rapidly decreasing on $\mathbb{H}^n = R^{2n+1}$. Note that we could equally specify that $X_1 \cdots X_K u$ be rapidly decreasing for any right invariant vector fields X_j. It is easy to see that $\mathcal{S}(\mathbb{H}^n)$ is a convolution algebra:

$$u,v \in \mathcal{S}(\mathbb{H}^n) \Rightarrow u*v \in \mathcal{S}(\mathbb{H}^n); \tag{1.34}$$

we are of course using Heisenberg group convolution. If $u \in \mathcal{S}(\mathbb{H}^n)$, then $\hat{u}(\tau,y,\eta) \in \mathcal{S}(R^{2n+1})$, and vice versa. Note that

$$\pi_{\pm\lambda}(u) = \hat{u}(\pm\lambda, \pm\lambda^{\frac{1}{2}}X, \lambda^{\frac{1}{2}}D)$$

is hence a rapidly decreasing function of λ with values in $OP\mathcal{S}_1^{-\infty}$ as $\lambda \to \infty$, where we define the class $\mathcal{S}_1^m$ to consist of smooth $a(x,\xi)$ such that

$$|D_x^\beta D_\xi^\alpha a(x,\xi)| \le C_{\alpha\beta}(1+|x|+|\xi|)^{m-|\alpha|-|\beta|}.$$

There is a slightly delicate matter of specifying whether u belongs to $\mathcal{S}(\mathbb{H}^n)$ purely in terms of

$$\pi_{\pm\lambda}(u) = \sigma_u(\pm\lambda)(X,D), \tag{1.35}$$

namely to specify adequately the behavior of $\pi_{\pm\lambda}(u)$ as $\lambda \to 0$. Of course, since

$$\hat{u}(\pm\lambda,y,\eta) = \sigma_u(\pm\lambda)(\pm\lambda^{-\frac{1}{2}}y, \lambda^{-\frac{1}{2}}\eta), \tag{1.36}$$

we could simply specify that the right side of (1.36) define an element of $\mathcal{S}(\mathbb{R}^{2n+1})$. We call the reader's attention to work of Geller [G3], describing $u \in \mathcal{S}(\mathbb{H}^n)$ via $\pi_{\pm\lambda}(u)$.

We now describe a certain transform of $u \in \mathcal{S}(\mathbb{H}^n)$, which struck the author as unexpected and amusing. This transform also helps to establish a certain technical point later in this section. To $u \in \mathcal{S}(\mathbb{H}^n)$, associate $\kappa(u)$ defined by inverting the parameter λ:

$$\sigma_{\kappa(u)}(\pm\lambda)(X,D) = \sigma_u(\pm\lambda^{-1})(X,D). \tag{1.37}$$

This is equivalent to saying

$$\widehat{\kappa(u)}(\pm\tau,y,\eta) = \hat{\ell}(\pm\tau,y,\eta) = \hat{u}(\pm\tau^{-1},\tau^{-1}y,\tau^{-1}\eta),$$

or

$$\hat{u}(\pm\lambda,y,\eta) = \hat{\ell}(\pm\lambda^{-1},\lambda^{-1}y,\lambda^{-1}\eta).$$

Thus it is clear that $\ell = \kappa(u)$ is a well defined element of $\mathcal{S}'(\mathbb{R}^{2n+1})$. Indeed, as we shall see, it is quite a special element. How special it is will be made clear by the following graphical representation of the transformation $(\tau,\bar{y},\bar{\eta}) = (\pm\lambda^{-1},\lambda^{-1}y,\lambda^{-1}\eta)$.

We set $z = (y,\eta)$, $\bar{z} = (\bar{y},\bar{\eta})$.

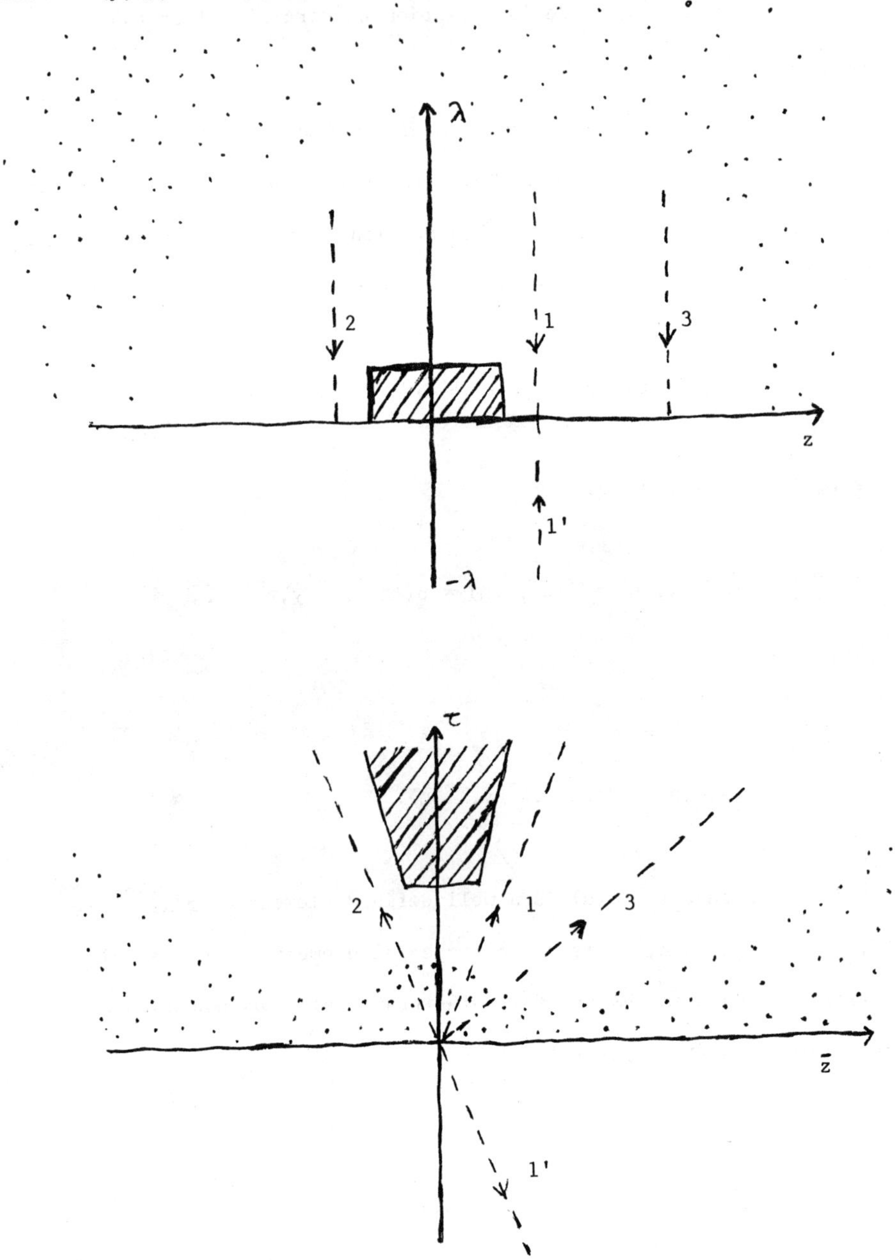

Figure 1.

If we first restrict attention to $\hat{u}$ on the upper half space $\lambda \geq 0$, we see that $\hat{u}$ coincides with the restriction to $\{\lambda \geq 0\}$ of an element of $\mathcal{S}(\mathbb{R}^{2n+1})$ if and only if $\hat{\ell}(\tau,y,\eta)$, restricted to $\tau \geq 0$, is equal to $\hat{\ell}_+(\tau,y,\eta)$, with the following three properties:

(1.38) $$\hat{\ell}_+(\tau,y,\eta) \in S^0_{1\#}(\mathbb{R}^{2n+1}),$$

(1.39) $$\hat{\ell}_+ \sim \sum_{j\geq 0} \hat{\ell}_j(\tau,y,\eta),$$

with $\hat{\ell}_j$ homogeneous of degree $-j$ in (τ,y,η), and

(1.40) $$\hat{\ell}_+(\tau,y,\eta) = 0 \text{ for } \tau < 0.$$

Note that (1.38) in particular says $\hat{\ell}_+(\tau,y,\eta)$ is smooth in all of $\mathbb{R}^{2n+1}$, so (1.40) implies $\hat{\ell}(\tau,y,\eta)$ vanishes to infinite order at $\tau = 0$, for $\tau \geq 0$. In addition we see that $\hat{u}$ on $\lambda \leq 0$ coincides with the restriction to $\{\lambda \leq 0\}$ of an element of $\mathcal{S}(\mathbb{R}^{2n+1})$ if and only if $\hat{\ell} = \hat{\ell}_-$ on $\tau < 0$, where $\hat{\ell}_-$ satisfies (1.38),(1.39), and, instead of (1.40), we have

(1.41) $$\hat{\ell}_-(\tau,y,\eta) = 0 \text{ for } \tau > 0.$$

To see whether $\hat{u}$ satisfying both these conditions is actually in $\mathcal{S}(\mathbb{R}^{2n+1})$ we need to know that all the appropriate compatibility conditions hold at $\lambda = 0$. It is simplest to state these if we break up $\hat{u}$ into its even and odd parts:

(1.42) $$\hat{u} = \hat{u}_e + \hat{u}_o,$$

(1.43) $$\hat{u}_e(\lambda,y,\eta) = \hat{u}_e(-\lambda,y,\eta), \quad \hat{u}_o(\lambda,y,\eta) = -\hat{u}_o(-\lambda,y,\eta).$$

Clearly the compatibility conditions are equivalent to

(1.44) $$D^j_\lambda \hat{u}_e(0,y,\eta) = 0 \text{ for } j \text{ odd}; \quad D^j_\lambda \hat{u}_o(0,y,\eta) = 0 \text{ for } j \text{ even}$$
(and $j=0$).

The associated decomposition

$$(1.45)\qquad \hat{\ell}(\tau,y,\eta) = \hat{\ell}_e(\tau,y,\eta) + \hat{\ell}_o(\tau,y,\eta)$$

stores up these conditions in the form

$$(1.46)\qquad \begin{aligned} \hat{\ell}_e(\tau,y,\eta) &\sim \sum_{j\geq 0} \hat{\ell}_{2j}(\tau,y,\eta) \\ \hat{\ell}_o(\tau,y,\eta) &\sim \sum_{j\geq 0} \hat{\ell}_{2j+1}(\tau,y,\eta) \end{aligned}$$

where $\hat{\ell}_k(\tau,y,\eta)$ is homogeneous of degree $-k$ in (τ,y,η). If we let $\kappa_1(u)$ denote the operator of left convolution by $\kappa(u)$, these observations product the:

<u>Inversion Trick.</u> The transformation $u \mapsto \kappa_1(u)$ is an isomorphism of $\mathcal{S}(\mathbb{H}^n)$ onto the class of right invariant pseudodifferential operators on $\mathbb{H}^n$, in OPS^0, $Lf = \ell * f$ ($\ell = \kappa(u)$) with $\hat{\ell}$ characterized by (1.38)-(1.41) and (1.45)-(1.46). If only (1.38)-(1.41) are considered, we get the isomorphic image of the "piecewise" elements $\hat{u}$ of $\mathcal{S}(\overline{\mathbb{R}_+^{2n+1}}) + \mathcal{S}(\overline{\mathbb{R}_-^{2n+1}})$. Furthermore, these isomorphisms are isomorphisms of convolution algebras:

$$(1.47)\qquad \kappa(u*v) = \kappa(u)*\kappa(v).$$

Note that (1.47) follows directly from (1.37). Note also that $\kappa(u) \in \mathcal{S}'(\mathbb{H}^n)$ is singular only at the origin (the identity element of $\mathbb{H}^n$) and is equal to an element of $\mathcal{S}(\mathbb{H}^n)$ outside any neighborhood of the origin. Let us denote by $OP\mathbb{C}$ the set of convolution operators $Lu = \ell * u$ with $\hat{\ell}$ satisfying (1.38)-(1.41) and $OP\mathbb{C}_0$ those which also satisfy (1.45)-(1.46). Thus $OP\mathbb{C}_0$ is the image of the convolution algebra $\mathcal{S}(\mathbb{H}^n)$. It follows that $OP\mathbb{C}_0$ is a convolution algebra. It is clear that $OP\mathbb{C}$ is a convolution algebra; hence the set of u with u a piecewise element of $\mathcal{S}(\mathbb{R}^{2n+1})$, with simple jumps across $\lambda = 0$, is also a convolution algebra; let us call this convolution algebra $\hat{\mathcal{S}}_p(\mathbb{H}^n)$.

Let us note that $\hat{u}(\tau,y,\eta) \in \mathcal{S}(\mathbb{R}^{2n+1})$ vanishes to infinite order at

$(\lambda,y,\eta) = 0$ if and only if $\hat{\ell}(\tau,y,\eta) \in S^0_{1\#}$ vanishes to infinite order along the rays $(\pm\tau,0,0)$, $\tau \to \infty$. We can restate this:

<u>Corollary to inversion trick</u>. The map $u \to \kappa_1(u)$ sets up an isomorphism between the set of $u \in \mathcal{S}(\mathbb{H}^n)$ such that $\hat{u}(\lambda,y,\eta)$ vanishes to infinite order at $(\lambda,y,\eta) = 0$ and the subalgebra of $OP\mathfrak{C}_0$ consisting of pseudo-differential operators whose full symbols vanish to infinnite order on the conic subset of $T^*\mathbb{H}^n \setminus 0$ which is the right invariant set whose fiber over the origin is generated by $\pm dt$.

This line bundle Λ, which is being intersected with $T^*\mathbb{H}^n \setminus 0$, furnishes a contact structure on $\mathbb{H}^n$, and will arise in other contexts later. We remark that the property of a pseudodifferential operator in $OPS^m_{1,0}(\Omega)$ (or even more general classes) of having its complete symbol vanish to infinite order on a closed conic subset of $T^*(\Omega) \setminus 0$ is invariant.

Let us record the following definition.

(1.48) $$u \in \mathcal{S}_{oo}(\mathbb{H}^n) \iff \hat{u}(\tau,y,\eta) \in \mathcal{S}(\mathbb{R}^{2n+1}) \text{ vanishes to infinite order at the origin.}$$

We will use the corollary, characterizing the image under κ of $\mathcal{S}_{oo}(\mathbb{H}^n)$, to study the action on $\mathcal{S}_{oo}(\mathbb{H}^n)$ of convolution operators $Lu = \ell * u$ where $\hat{\ell}$ is singular at the origin, so L does not map $\mathcal{S}(\mathbb{H}^n)$ into itself. This will be useful in streamlining some results to be developed in the next section. We begin by observing that $u \in \mathcal{S}_{oo}(\mathbb{H}^n)$ if and only if $u \in \mathcal{S}(\mathbb{H}^n)$ and $\langle u,p\rangle = 0$ for every polynomial function p on $\mathbb{H}^n = \mathbb{R}^{2n+1}$. Since the translate $p_y(x) = p(y^{-1}x)$ $(x,y \in \mathbb{H}^n)$ is a polynomial in x whenever $p(x)$ is a polynomial, we see that if $u \in \mathcal{S}_{oo}(\mathbb{H}^n)$, then so is $u_y(x) = u(y^{-1}x)$. This easily gives:

(1.49) $$\ell * u \in \mathcal{S}_{oo}(\mathbb{H}^n) \text{ if } u \in \mathcal{S}_{oo}(\mathbb{H}^n),\ \ell \in \mathcal{E}'(\mathbb{H}^n),$$

and

(1.50) $\ell * u \in \mathcal{S}_{oo}(\mathbb{H}^n)$ if $u \in \mathcal{S}_{oo}(\mathbb{H}^n)$, $\ell \in \mathcal{S}(\mathbb{H}^n)$.

Of course, (1.50) can also be obtained from the corollary to the inversion trick, since $L_j \in OP\mathfrak{C}_0$ implies that $L_1L_2 \in OP\mathfrak{C}_0$ has full symbol vanishing to infinite order at Λ provided either factor does.

We now want to study $Ku = k*u$ for $u \in \mathcal{S}_{oo}(\mathbb{H}^n)$, where $\hat{k}(\tau,y,\eta)$ satisfies the following conditions:

(1.51) $\hat{k} \in C^\infty(\mathbb{R}^{2n+1} \setminus 0)$,

(1.52) $\hat{k}(\pm s\lambda, s^{\frac{1}{2}}y, s^{\frac{1}{2}}\eta) = s^{m/2}\hat{k}(\pm\lambda,y,\eta), \quad s > 0.$

Note that (1.52) implies

(1.53) $\sigma_K(\pm\lambda)(X,D) = \lambda^{m/2}\sigma_K(\pm 1)(X,D).$

Now if $\chi(y,\eta)$ is a smooth cut-off, equal to 1 for $|y|+|\eta| \le \sigma$, 0 for $|y|+|\eta| \ge 2\sigma$, and if $\psi(\lambda)$ is a smooth cut-off, equal to 1 for $|\lambda| \ge \sigma$, 0 for $|\lambda| \ge 2\sigma$, write

$$\hat{k} = \hat{k}_1 + \hat{k}_2 = \chi\psi\hat{k} + (1-\chi\psi)\hat{k}.$$

Thus $\hat{k}_1$ is supported on $|\lambda| \le 2\sigma$, $|y|+|\eta| \le 2\sigma$ while $\hat{k}_2 \in S^m_{\frac{1}{2}\#}$ vanishes near 0. It follows that k_2 is the sum of an element of $\mathcal{S}(\mathbb{H}^n)$ and an element of $\mathcal{E}'(\mathbb{H}^n)$, so, by (1.49) and (1.50), $k_2 * u \in \mathcal{S}_{oo}(\mathbb{H}^n)$ if $u \in \mathcal{S}_{oo}(\mathbb{H}^n)$. Now we look at $k_1 * u = K_1 u$, for $u \in \mathcal{S}_{oo}(\mathbb{H}^n)$. Note that, if

$$\ell = \kappa(k_1 * u),$$

we have, for $Lf = \ell * f$,

$$L \in OPS^0(\mathbb{H}^n),$$

since the full symbol of $\kappa_1(u)$ vanishes to infinite order at Λ. Hence,

$$L \in OP\mathfrak{C}.$$

This implies that k_1*u has Euclidean Fourier transform piecewise in $\mathcal{S}$ on $\overline{\mathbb{R}^{2n+1}_+}$ and $\overline{\mathbb{R}^{2n+1}_-}$. In fact, the full symbol of L vanishes outside a small conic neighborhood of Λ, if σ is chosen small, and this implies the Fourier transform of k_1*u vanishes to infinite order as $\pm\tau \to 0$, for $|y|+|\eta| \geq \sigma'$, where σ' is small if σ is. It follows that $Ku = k_1*u + k_2*u$ has Euclidean Fourier transform piecewise in $\mathcal{S}$ in $\overline{\mathbb{R}^{2n+1}_+}$ and $\overline{\mathbb{R}^{2n+1}_-}$, and these functions in the upper and lower half spaces match up at $\tau = 0$ to be smooth, provided $(y,\eta) \neq (0,0)$. But the origin cannot carry any singularity, so in fact $Ku \in \mathcal{S}(\mathbb{H}^n)$. Furthermore, since the full symbol of $L = \kappa_1(k_1*u)$ vanishes to infinite order at Λ, it follows that $k_1*u \in \mathcal{S}_{oo}(\mathbb{H}^n)$, so finally we have

(1.53) $\qquad k*u \in \mathcal{S}_{oo}(\mathbb{H}^n)$ if $u \in \mathcal{S}_{oo}(\mathbb{H}^n)$ and k satisfies (1.51)-(1.52).

Let us renotate the decomposition $k = k_1 + k_2$ by

(1.54) $$k = k^o + k^b.$$

Here, k^o is singular only at the origin, and equal to an element of $\mathcal{S}(\mathbb{H}^n)$ outside a neighborhood of the origin, while k^b is C^∞, with $\hat{k}^b(\tau,y,\eta)$ supported in $|\tau| \leq 2\sigma$, $|y|+|\eta| \leq 2\sigma$. If k_j satisfy (1.51)-(1.52) with decompositions $k_j = k_j^o + k_j^b$ as in (1.54), set $K_jf = K_j^of + K_j^bf = k_j^o*f + k_j^b*f$. Note that $K_j^o: \mathcal{S}(\mathbb{H}^n) \to \mathcal{S}(\mathbb{H}^n)$ and $K_j^b: \mathcal{E}'(\mathbb{H}^n) \to C^\infty(\mathbb{H}^n)$. Now

(1.55) $$K_3(u - \chi u) = k_1^o*k_2^o*(u - \chi u) + k_1^o*k_2^b*(u - \chi u) + k_1^b*k_2^o*(u - \chi u) + k_1^b*k_2^b*(u - \chi u)$$

where K_3 is defined by

(1.56) $$\sigma_{K_3}(\pm\lambda)(X,D) = \sigma_{K_1}(\pm\lambda)(X,D)\sigma_{K_2}(\pm\lambda)(X,D),$$

and χu is given by

(1.57) $$\chi u = \psi*u$$

where $\hat{\psi}(\tau,y,\eta)$ is a smooth function, with compact support, equal to 1 for $|\tau| \le 2\sigma$, $|y|+|\eta| \le 2\sigma$. Note that the second and third terms on the right side of (1.55) represent right convolution of $u-\chi u$ by C^∞ functions. This identity generalizes to more general distributions, in particular to $u = \delta$. We have

$$K_3(\delta-\chi\delta) = K_1^o K_2^o \delta + k_1^b * k_2^b(\delta-\chi\delta) + \varphi, \quad \varphi \in C^\infty. \tag{1.58}$$

Now, under the transformation κ_1, we get from $k_1^b * k_2^b *(\delta-\chi\delta)$ the operator product of a pair of right invariant pseudodifferential operators with symbol supported in a certain conic neighborhood of Λ, with a right invariant pseudodifferential operator whose full symbol vanishes on a conic neighborhood of such support. Such a product is a smoothing operator, and hence belongs to the image under κ_1 of an element of $\mathcal{S}(\mathbb{R}^{2n+1})$. In other words,

$$K_3\delta = K_1^o K_2^o \delta + \tilde{\varphi}, \quad \tilde{\varphi} \in C^\infty(\mathbb{H}^n). \tag{1.59}$$

This enables one to manipulate some classes of pseudodifferential operators symbolically, in a clean fashion, without worrying too much about the necessity to smooth out the singularity of their symbols at the origin. Compare the remarks at the end of the proof of proposition 2.6.

2. Right invariant pseudodifferential operators on $\mathbb{H}^n$.

Having made a basic study of convolution operators on $\mathbb{H}^n$, we now determine some classes of operators that deserve the exalted title of pseudodifferential operators. One impartant class is generated by the study of the "Heisenberg Laplacian"

$$\mathcal{L}_0 = \sum_{j=1}^{n} (L_j^2 + M_j^2), \tag{2.1}$$

a second order nonelliptic (but hypoelliptic) operator. Note that, by (1.18), (1.19),

$$\pi_{\pm\lambda}(\mathcal{L}_0) = \lambda \sum_{j=1}^{n} (\partial^2/\partial x_j^2 - x_j^2) \tag{2.2}$$

and

$$\pi_{(y,\eta)}(\mathcal{L}_0) = \sum_{j=1}^{n} (y_j^2 + \eta_j^2). \tag{2.3}$$

We define the following classes:

Definition 2.1. The class Ψ_0^m consists of functions $\hat{k}(\tau,y,\eta)$, smooth except at 0, such that, for $\tau > 0$,

$$\hat{k}(\pm\tau,y,\eta) = \tau^{m/2}\hat{k}(\pm 1,\tau^{-\frac{1}{2}}y,\tau^{-\frac{1}{2}}\eta). \tag{2.4}$$

Neglecting the singularity at the origin, elements of Ψ_0^m belong to $S^m_{\frac{1}{2}\#}$ if $m \geq 0$ and to $S^{m/2}_{\frac{1}{2}\#}$ if $m < 0$. If $\hat{k} \in \Psi_0^m$, we say $K \in OP\Psi_0^m$, if $Ku = k*u$. Proposition 1.1 of Chapter I implies $OP\Psi_0^m \subset OPS^m_{\frac{1}{2},\frac{1}{2}}$ for $m \geq 0$, $OPS^{m/2}_{\frac{1}{2},\frac{1}{2}}$ for $m < 0$. In fact, using the dilations $\alpha(s)(t,q,p) = (\sigma^2 t,\sigma q,\sigma p)$, $\sigma = e^s$, we see that, modulo smoothing operators, $OP\Psi_0^m = OP\mathcal{H}(\mathbb{H}^n,\alpha,m)$, a special case of a class introduced in Chapter I. From (1.15), we have $K \in OP\Psi_0^m$ provided $\hat{k}(\tau,y,\eta)$ is smooth away from $(\tau,y,\eta) = 0$ and, for $\lambda > 0$,

$$\sigma_K(\pm\lambda)(x,\xi) = \lambda^{m/2}\,\sigma_K(\pm 1)(x,\xi). \tag{2.5}$$

Thus (2.2) gives

$$\mathcal{L}_0 \in OP\Psi_0^2. \tag{2.6}$$

Note also that

$$\pi_{\pm\lambda}(T) = \pm i\lambda \tag{2.7}$$

and hence

$$T \in OP\Psi_0^2. \tag{2.8}$$

The following is a convenient characterization of Ψ_0^m. For a related result, see Geller [G2].

Proposition 2.2. Let $a_\pm(x,\xi) \in C^\infty(\mathbb{R}^{2n})$. Then (2.5) defines an element of Ψ_o^m, with $\sigma_K(\pm1)(x,\xi) = a_\pm(x,\xi)$, if and only if $a_\pm(x,\xi)$ have the compatible asymptotic expansions

$$a_\pm(ry,r\eta) \sim \sum_{j\geq 0} r^{m-2j} (\pm1)^j \varphi_j(\pm y,\eta) \text{ as } r \to +\infty, \tag{2.9}$$

for $|y|^2+|\eta|^2 = 1$, $\varphi_j \in C^\infty(S^{2n-1})$.

Proof. We have

$$\hat{k}(\pm\tau,y,\eta) = \tau^{m/2} a_\pm(\pm\tau^{-\frac{1}{2}},\tau^{-\frac{1}{2}}\eta),$$

for $\tau \neq 0$. The only problem is to specify when this extends to a function smooth at $\tau = 0$, $(y,\eta) \neq (0,0)$. Writing

$$\hat{k}(\pm\tau,y,\eta) \sim \sum_{j\geq 0} (\pm\tau)^j \varphi_j(y,\eta)$$

$$= \tau^{m/2} a_\pm(\pm\tau^{-\frac{1}{2}}y,\tau^{-\frac{1}{2}}\eta)$$

and changing variables, setting $r = \tau^{-\frac{1}{2}}$, makes the characterization (2.9) apparent.

A characterization equivalent to (2.9) is

$$\sigma_K(\pm1)(x,\xi) \sim \sum_{j\geq 0} (\pm1)^j \varphi_j(\pm x,\xi) \quad \text{as } |x|^2 + |\xi|^2 \to \infty, \tag{2.10}$$

with $\varphi_j(x,\xi)$ homogeneous of degree m-2j in (x,ξ). Note from (1.17) that

$$\pi_{(y,\eta)}(k) = \varphi_0(y,\eta). \tag{2.11}$$

In other words, $\pi_{(y,\eta)}(k)$ is the principal symbol of the operator $\sigma_K(+1)(X,D)$, evaluated at (y,η).

Note in particular that if $\sigma_K(\pm1)(x,\xi) \in \mathcal{S}(\mathbb{R}^{2n})$, the Schwartz space of rapidly decreasing functions, then (2.5) defines an element of $OP\Psi_o^m$.

Since $\pi_{\pm1}(\mathcal{L}_0+i\alpha T) = \sum_1^n(\partial^2/\partial x_j^2 - x_j^2)\mp\alpha$, we have

(2.12) $$\sigma_{\mathcal{L}_0+i\alpha T}(\underline{+}1)(x,\xi) = -|\xi|^2 - |x|^2 \mp \alpha,$$

which of course is consistent with (2.10).

Proposition 2.2 motivates us to make the following definitions.

Definition 2.3. We say $a(x,\xi) \in \mathcal{H}_b^m$ if $a(x,\xi)$ is smooth and has the asymptotic expansion

(2.13) $$a(x,\xi) \sim \Sigma\varphi_j(x,\xi) \qquad \text{as } |x|^2 + |\xi|^2 \to \infty,$$

where $\varphi_j(x,\xi)$ is homogeneous of degree $m-2j$ in (x,ξ).

Definition 2.4. We say the pair $a_{\underline{+}}(x,\xi)$ belongs to $\mathcal{H}^m$ if both $a_+(x,\xi)$ and $a_-(x,\xi)$ belong to $\mathcal{H}_b^m$, and if furthermore their expansions are compatible, in the sense that

(2.14) $$a_{\underline{+}}(x,\xi) \sim \sum_{j\geq 0} (\underline{+}1)^j \varphi_j(\underline{+}x,\xi).$$

The content of proposition 2.2 is hence that $K \in OP\Psi_o^m$ precisely when (2.5) holds and $\sigma_K(\underline{+}1)(x,\xi) \in \mathcal{H}^m$. Note that if $L \in OP\Psi_o^\mu$, then

(2.15) $$\sigma_{KL}(\underline{+}\lambda)(X,D) = \lambda^{(m+\mu)/2}\sigma_K(\underline{+}1)(X,D)\sigma_L(\underline{+}1)(X,D), \qquad \lambda > 0.$$

Thus, to deduce from proposition 2.2 that the composition KL belongs to $OP\Psi_o^{m+\mu}$, we need to discuss compositions of operators in the Weyl calculus. This theory has been worked out, in enough generality for our present needs, in [G11], and in much greater generality, in [H10]. We state some results here, referring to these sources for proofs.

The class $\mathcal{H}_b^m$ of symbols is a subset of the class $\mathcal{S}_1^m$, which we define to consist of $a(x,\xi)$, smooth, such that

(2.16) $$|D_x^\beta D_\xi^\alpha\, a(x,\xi)| \leq C_{\alpha\beta}(1+|x|+|\xi|)^{m-|\alpha|-|\beta|}.$$

If $a(x,\xi) \in \mathcal{S}_1^m$, we say $a(X,D) \in OP\mathcal{S}_1^m$; similarly if $a(x,\xi) \in \mathcal{H}_b^m$, we say $a(X,D) \in OP\mathcal{H}_b^m$; finally if $a_{\underline{+}}(x,\xi) \in \mathcal{H}^m$, we say $a_{\underline{+}}(X,D) \in OP\mathcal{H}^m$. Now if

$a(X,D) \in OP\mathcal{S}_1^m$ and $b(X,D) \in OP\mathcal{S}_1^\mu$, then

$$a(X,D)b(X,D) = c(X,D) \in OP\mathcal{S}_1^{m+\mu}, \tag{2.17}$$

and $c(x,\xi)$ has the asymptotic expansion

$$c(x,\xi) \sim \sum_{j\geq 0} (1/j!)\ \{a,b\}_j(x,\xi) \tag{2.18}$$

where $\{a,b\}_j(x,\xi)$ is defined by

$$\{a,b\}_0(x,\xi) = a(x,\xi)b(x,\xi), \tag{2.19}$$

and for $j \geq 1$,

$$\{a,b\}_j(x,\xi) = (-i/2)^j \cdot \sum_{k=1}^{n} (\partial^2/\partial y_k\partial\xi_k - \partial^2/\partial x_k\partial\eta_k)^j a(x,\xi)b(y,\eta)\Big|_{\substack{y=x\\ \eta=\xi}}. \tag{2.20}$$

Note that if $a_k(x,\xi)$ and $b_\ell(x,\xi)$ are homogeneous of degree k and ℓ, respectively (not necessarily integers), then $\{a_k,b_\ell\}_j(x,\xi)$ is homogeneous of degree $k+\ell-2j$ in (x,ξ). This shows that if $a(X,D) \in OP\mathcal{H}_b^m$ and $b(X,D) \in OP\mathcal{H}_b^\mu$, then $a(X,D)b(X,D) \in OP\mathcal{H}_b^{m+\mu}$. Furthermore, we have:

<u>Proposition 2.5</u>. If $a_\pm(X,D) \in OP\mathcal{H}^m$ and $b_\pm(X,D) \in OP\mathcal{H}^\mu$, then

$$a_\pm(X,D)b_\pm(X,D) = c_\pm(X,D) \in OP\mathcal{H}^{m+\mu}. \tag{2.21}$$

<u>Proof</u>. We know that $c_+(X,D)$ and $c_-(X,D)$ belong to $OP\mathcal{H}_b^{m+\mu}$ and

$$c_\pm(x,\xi) \sim \sum_{j\geq 0} (1/j!)\{a_\pm,b_\pm\}_j(x,\xi). \tag{2.22}$$

Let us say

$$f_\pm(x,\xi) \in \mathcal{H}^{m,k} \text{ if } f_\pm(x,\xi) \sim \sum_{j\geq k} (\pm 1)^j \psi_j(\pm x,\xi) \tag{2.23}$$

with $\psi_j(x,\xi)$ homogeneous of degree $m-2j$ in (x,ξ). Note that $\mathcal{H}^{m,k} \subset \mathcal{H}^m \cap \mathcal{S}_1^{m-2k}$, and formal series of elements of $\mathcal{H}^{m,k}$, $k=1,2,\ldots$, asymptotically sum to elements of $\mathcal{H}^m$. Now the jth term of (2.22) is seen to belong to $\mathcal{H}^{m+\mu,j}$, so the proof is complete.

An immediate consequence of proposition 2.5 is:

Proposition 2.6. If $K_1 \in OP\Psi_o^m$, $K_2 \in OP\Psi_o^\mu$, then $K_1K_2 \in OP\Psi_o^{m+\mu}$, and

$$\sigma_{K_1K_2}(\underline{+}\lambda)(X,D) = \sigma_{K_1}(\underline{+}\lambda)(X,D)\sigma_{K_2}(\underline{+}\lambda)(X,D). \tag{2.24}$$

In view of the discussion at the end of section 1, leading up to (1.59), we have a similar result for the operators obtained by regularizing $\hat{k}_j$ near the origin.

Similarly we can discuss adjoints. It follows from the Weyl calculus that if $a(X,D) \in OP\mathcal{S}_1^m$, then $a(X,D)^* = a^*(X,D) \in OP\mathcal{S}_1^m$, and

$$a^*(x,\xi) = \overline{a(x,\xi)}. \tag{2.25}$$

This formula shows that if $a(X,D) \in OP\mathcal{H}_b^m$ then $a(X,D)^* \in OP\mathcal{H}_b^m$ and furthermore if $a_{\underline{+}}(X,D) \in OP\mathcal{H}^m$ then $a_{\underline{+}}(X,D)^* \in OP\mathcal{H}^m$. Consequently we have

Proposition 2.7. If $K \in OP\Psi_o^m$, then $K^* \in OP\Psi_o^m$, and

$$\sigma_{K^*}(\underline{+}\lambda)(X,D) = \sigma_K(\underline{+}\lambda)(X,D)^*. \tag{2.26}$$

Our next task is to examine $K \in OP\Psi_o^m$ in the case when the operators $\sigma_K(\underline{+}1)(X,D)$ are elliptic. In general, whenever $a(X,D) \in OP\mathcal{S}_1^m$ is elliptic, there exists a parametrix $b(X,D) \in OP\mathcal{S}_1^{-m}$.

Proposition 2.8. If $a_{\underline{+}}(X,D) \in OP\mathcal{H}^m$ with both operators elliptic, having parametrices $b_{\underline{+}}(X,D)$ (belonging a priori to $OP\mathcal{S}_1^{-m}$), then $b_{\underline{+}}(X,D) \in OP\mathcal{H}^{-m}$.

Proof. The content of this proposition is that

$$b_{\underline{+}}(x,\xi) \sim \sum_{j\geq 0} (\underline{+}1)^j \psi_j(\underline{+}x,\xi) \tag{2.27}$$

with $\psi_j(\underline{+}x,\xi)$ homogeneous of degree $-m-2j$ in (x,ξ). If we assume (2.14), pick $\beta_{\underline{+}}(x,\xi)$ smooth with

$$\beta_{\underline{+}}(x,\xi) = \varphi_0(\underline{+}x,\xi)^{-1} \qquad \text{for } |x|^2 + |\xi|^2 \text{ large}. \tag{2.28}$$

Then clearly $\beta_{\pm}(x,\xi) \in \mathcal{H}^{-m}$. By (2.17),(2.18), we have

$$\beta_{\pm}(X,D)a_{\pm}(X,D) = I + r_{\pm}(X,D) \tag{2.29}$$

with

$$r_{\pm}(x,\xi) \in \mathcal{H}^{0,1}. \tag{2.30}$$

A simple extension of proposition 2.5 is that

$$r_j^{\pm}(X,D) \in OP\mathcal{H}^{m_j,k_j} \Rightarrow r_1^{\pm}(X,D)r_2^{\pm}(X,D) \in OP\mathcal{H}^{m_1+m_2,k_1+k_2}. \tag{2.31}$$

In particular,

$$r_{\pm}(X,D)^k \in OP\mathcal{H}^{0,k} \subset OP\mathcal{S}_1^{-2k}. \tag{2.32}$$

Consequently

$$b_{\pm}(X,D) \sim (I - r_{\pm}(X,D) + r_{\pm}(X,D)^2 - \cdots)\beta_{\pm}(X,D) \tag{2.33}$$

belongs to $OP\mathcal{H}^{-m}$, as asserted.

Suppose that $K \in OP\Psi_o^m$ and that $\sigma_K(\pm 1)(X,D)$ are elliptic. Then denoting parametrices by $\sigma_L(\pm 1)(X,D)$, we obtain an operator $L \in OP\Psi_o^{-m}$. It does not follow that $KL - I$ or $LK - I$ is smoothing. Rather, one has

$$\sigma_{KL}(\pm\lambda)(X,D) = I + r_{\pm}(X,D) \tag{2.34}$$

with $r_{\pm}(x,\xi) \in \mathcal{S}_1^{-\infty}$. An operator $R \in OP\Psi_o^0$ with

$$\sigma_R(\pm\lambda)(x,\xi) \in \mathcal{S}_1^{-\infty} \tag{2.35}$$

will be said to belong to $OP\Psi_o^{0,\infty}$. More generally, we have the following notion:

<u>Definition 2.9</u>. We say $R \in OP\Psi_o^{m,k}$ if $R \in OP\Psi_o^m$ with

$$\sigma_R(\pm 1)(X,D) \in OP\mathcal{H}^{m,k}. \tag{2.36}$$

We denote $\bigcap_{k\geq 0} OP\Psi_o^{m,k}$ by $OP\Psi_o^{m,\infty}$.

In order to construct a (left of right) parametrix for $K \in OP\Psi_o^m$ with $\sigma_K(\pm 1)$ elliptic, it is necessary to assume $\sigma_K(\pm 1)(X,D)$ has a (left or right) inverse. It is worth noting that, from [B4], or [G12], the following three conditions are equivalent if $a(X,D) \in OP\mathcal{S}_1^m$ is elliptic:

(2.37) $a(X,D)$ has a left inverse in $OP\mathcal{S}_1^{-m}$.

(2.38) $a(X,D)$ is injective on the Schwartz space $\mathcal{S}(\mathbb{R}^n)$,

(2.39) $a(X,D)$ is injective on the space $\mathcal{S}'(\mathbb{R}^n)$ of tempered distributions.

Similarly, the following three conditions are equivalent:

(2.40) $a(X,D)$ has a right inverse in $OP\mathcal{S}_1^{-m}$,

(2.41) $a(X,D)$ is surjective on the Schwartz space $\mathcal{S}(\mathbb{R}^n)$,

(2.42) $a(X,D)^*$ is injective on $\mathcal{S}(\mathbb{R}^n)$.

Our regularity result is the following.

Theorem 2.10. If $K \in OP\Psi_o^m$ has the property that $\sigma_K(\pm 1)(X,D)$ are elliptic, then K has a left inverse $L \in OP\Psi_o^{-m}$ if and only if $\sigma_K(\pm 1)(X,D)^*$ are injective on $\mathcal{S}(\mathbb{R}^n)$, and such a right inverse if and only if $\sigma_K(\pm 1)(X,D)$ are injective in $\mathcal{S}(\mathbb{R}^n)$.

Proof. In light of the discussion above, this is an immediate consequence of proposition 2.8.

Corollary 2.11. If $K \in OP\Psi_o^m$ has the property that $\sigma_K(\pm 1)(X,D)$ are elliptic and injective on $\mathcal{S}(\mathbb{R}^n)$, then K is hypoelliptic.

In the case of scalar differential operators, the hypoellipticity is proved by Miller [M10], using somewhat different arguments; see also Rockland [R2]. A more general result has been proved by Helffer and Nourrigat [H3], by different menas. This result was also announced by Dynin [D2]; his announcement did not include proofs, but it is fairly likely that his argument was similar to that given here. If K is scalar, with $\sigma_K(\pm 1)(X,D)$ elliptic, then, as observed by Grusin [G12], the operators $\sigma_K(\pm 1)(X,D)$ are Fredholm of index zero, so one sided and two sided invertibility coincide. The propositions above also work for $k \times k$ matrices of operators, in which case $\sigma_K(\pm 1)(X,D)$ may have nonzero index. As a counterpoint to corollary 2.11, we have:

Proposition 2.12. If $K \in OP\Psi_0^m$ has the property that $\sigma_K(\pm 1)(X,D)$ are elliptic but not both injective on $\mathcal{S}(\mathbb{R}^n)$, then K is not hypoelliptic.

Proof. Say $\sigma_K(\pm 1)(X,D)$ has a nontrivial kernel. Such a null space must be a finite dimensional subspace of $\mathcal{S}(\mathbb{R}^n)$, so the orthogonal projection onto this null space is $r(X,D)$ with $r(x,\xi) \in \mathcal{S}_1^{-\infty}$. Then

$$\text{(2.43)} \qquad \begin{aligned} \sigma_S(+\lambda)(X,D) &= r(X,D) \\ & & (\lambda > 0) \\ \sigma_S(-\lambda)(X,D) &= 0 \end{aligned}$$

defines an element $S \in OP\Psi_0^{0,\infty}$, and $KS = 0$. Since S is right invariant and not smoothing, $v = S\delta_0$ is not smooth; it is smooth except at the origin. Since $Kv = 0$, K is not hypoelliptic in any neighborhood of the origin. By right invariance, K is not hypoelliptic on any open set.

Let us apply these results on hypoellipticity to the well known important example $\mathcal{L}_0+i\alpha T$, first treated by Folland and Stein [F4], by a different method. By (2.12), we have

$$\text{(2.44)} \qquad \sigma_{\mathcal{L}_0+i\alpha T}(\pm 1)(X,D) = \Delta - |x|^2 \mp \alpha,$$

which clearly gives an elliptic pair in $OP\mathcal{H}^2$. We determine the spectrum of $\Delta - |x|^2$. In fact, as is well known, $-d^2/dx^2 + x^2$, acting on functions on $\mathbb{R}$, has discrete spectrum consisting of the eigenvalue $2j+1$, $j=0,1,2,\ldots$, all simple. Indeed, the complete set of eigenfunctions of d^2/dx^2-x^2 is given by the set of Hermite functions

$$h_j(x) = [\pi^{\frac{1}{2}}2^j j!)]^{-\frac{1}{2}} (d/dx - x)^j e^{-x^2/2},$$

(2.45)

$$(d^2/dx^2-x^2)h_j(x) = -(2j+1)h_j(x).$$

It follows that, on $L^2(\mathbb{R}^n)$, $\Delta - |x|^2$ has spectrum consisting of all the negative integers of the form $-n-2k$, $k=0,1,2,\ldots$ Thus theorem 2.10 and proposition 2.12 yield the result of Folland and Stein:

Proposition 2.13. For $\alpha \in \mathbb{C}$, $\mathcal{L}_0+i\alpha T$ is hypoelliptic on $\mathbb{H}^n$ if and only if

(2.46) $$-(n\pm\alpha) \notin \{0,1,2,\ldots\}.$$

If (2.46) holds, $\mathcal{L}_0+i\alpha T$ has a two-sided inverse in $OP\Psi_o^{-2}$. If (2.46) fails, there exists a non-smoothing element S of $OP\Psi_o^{0,\infty}$ such that $(\mathcal{L}_0+i\alpha T)S = 0$.

For a large class of homogeneous second order polynomials $Q(x,\xi)$, the spectrum of $Q(X,D)$ is completely specified by the following result; see Sjöstrand [S6], Hörmander [H8], Grigis [G10].

Proposition 2.14. Let $Q(x,\xi)$ be a second order homogeneous polynomial in (x,ξ). Suppose that Q takes values in a cone of the form

(2.47) $$\Gamma = \{z \in \mathbb{C}: |\text{Im } z| \le K \text{ Re } z\}.$$

Denote by μ_j the eigenvalues in $\Gamma \setminus 0$ of the Hamilton map F of $i^{-1}Q$, defined by

(2.48) $$(1/i)\mathbb{Q}(v,v') = \sigma(v,Fv')$$

where σ is the symplectic form on $\mathbb{R}^{2n}$:

(2.49) $$\sigma((x,\xi),(x',\xi')) = x'\cdot\xi - x\cdot\xi',$$

and $\mathbb{Q}$ is the bilinear form polarizing Q. Let $N_o \subset \mathbb{C}^{2n}$ be the space of generalized eigenvectors of F associated to eigenvalue 0. Then α is not in the spectrum of Q(X,D) if and only if

$$(2.50)\qquad -\alpha + \mathbb{Q}(v,\bar{v}) + \sum_j (2k_j+1)\mu_j \neq 0 \text{ for all } v \in N_o,\ k_j \in \mathbb{Z}^+ \cup \{0\}.$$

In particular, if $Q(x,\xi)$ is positive definite, the spectrum of Q(X,D) consists of

$$(2.51)\qquad \{\sum_j (2k_j+1)\mu_j : k_j \in \mathbb{Z}^+ \cup \{0\}\}.$$

The result on the spectrum of Q(X,D) in the positive definite case is older; using the metaplectic representation, one can reduce it to a simple result in linear algebra. For an exposition of this case, see Chapter I of [T5]. For certain homogeneous polynomials of degree 3, the spectrum of Q(X,D) has been analyzed by Helffer [H1]. For general $Q(X,D) \in OP\mathcal{H}^m_b$, it would be out of the question to explicitly describe the spectrum, though qualitative studies of the spectrum have been made; see [H2] and references therein.

Here we mean the operator Q(X,D) to be defined by the Weyl calculus, of course. In particular, if $Q(x,\xi) = x_j\xi_k$, then $Q(X,D)u(x) = \frac{1}{2}i(x_j\partial u/\partial x_k + \partial(x_ju)/\partial x_k)$.

The fact that hypoellipticity of an element K of $OP\Psi^m_o$ depends not just on the ellipticity of $\sigma_K(\pm 1)(X,D)$ but also on the invertibility of these operators, is closely related to the essential non-commutativity of $OP\Psi^0_o$. In fact, if $L \in OP\Psi^\mu_o$, then the commutator $[K,L] = KL - LK$ has symbol

$$(2.52)\qquad \sigma_{[K,L]}(\pm\lambda)(X,D) = \lambda^{(m+\mu)/2}[\sigma_K(\pm 1)(X,D),\ \sigma_L(\pm 1)(X,D)].$$

Clearly this is not the symbol of an operator belonging to $OP\Psi^{m+\mu-1}_o$. In fact, the result one has is the following (recall Definition 2.9):

<u>Proposition 2.15</u>. If $K \in OP\Psi^m_o$, $L \in OP\Psi^\mu_o$, then

(2.53) $$[K,L] \in OP\Psi_o^{m+\mu,1}.$$

More generally, if $K \in OP\Psi_o^{m,k}$, $L \in OP\Psi_o^{\mu,\ell}$, then

(2.54) $$[K,L] \in OP\Psi_o^{m+\mu,k+\ell+1}.$$

Proof. It suffices to show that, if $a_{\pm}(X,D) \in OP\mathcal{H}^{m,k}$ and $b_{\pm}(X,D) \in OP\mathcal{H}^{\mu,\ell}$, then

(2.55) $$[a_{\pm}(X,D),b_{\pm}(X,D)] \in OP\mathcal{H}^{m+\mu,k+\ell+1}.$$

This follows from the formula (2.18) for the symbol of a product, along the same lines as the proof of proposition 2.5.

We now want to pass from operators in $OP\Psi_o^m$ with homogeneous symbols to nonhomogeneous operators given as asymptotic sums. We define the class $OP\Psi^m$ as follows.

Definition 2.16. We say a convolution operator K belongs to $OP\Psi^m$ if

(2.56) $$K \sim \sum_{j\geq 0} K_j, \quad K_j \in OP\Psi_o^{m-j},$$

where (2.56) means $K - \sum_{j=0}^{N} K_j$ is arbitrarily smoothing for N picked sufficiently large.

In order to fit $OP\Psi^m$ into the framework of Chapter I, we make Ψ^m into a Frechet space as follows. We say $\hat{k}(\tau,y,\eta) \in \Psi^m$ if $\hat{k}$ is C^∞ and

(2.57) $$\hat{k}(\tau,y,\eta) \sim \sum_{j\geq 0} k_j(\tau,y,\eta) \quad \text{as } |\tau|^2 + |y|^2 + |\eta|^2 \to \infty$$

where $\hat{k}_j$ satisfy the homogeneity conditions

(2.58) $$\hat{k}_j(r\tau,r^{\frac{1}{2}}y,r^{\frac{1}{2}}\eta) = r^{(m-j)/2}\,\hat{k}_j(\tau,y,\eta).$$

Here, (2.57) holds in the sense that, if $\varphi(\tau,y,\eta)$ vanishes near 0, is 1 near ∞,

$$\hat{k} - \varphi\sum_{1}^{N}\hat{k}_j \in \mathcal{S}_{\frac{1}{2}\#}^{m-N/2}.$$

Since we have required $\hat{k}$ to be C^∞, even at the origin, strictly speaking Ψ_0^m is not contained in Ψ^m, but altering $\hat{k}$ on a compact set in (τ,y,η) space changes a convolution operator by a smoothing operator, so, modulo smoothing operators, $OP\Psi_0^m$ is contained in $OP\Psi^m$.

Now to describe the Frechet space structure on Ψ^m, we first pick a large compact neighborhood of the origin and use the C^j norms of $\hat{k}$ on this neighborhood $\mathcal{O}$. For the rest, take $(\tau,y,\eta) \in S^{2n}$, and define

$$\kappa(s,\tau,y,\eta) = s^m\hat{k}(s^{-2}\tau, s^{-1}y, s^{-1}\eta), \quad s \leq 1. \tag{2.59}$$

Then $\hat{k} \in \Psi^m$ if and only if $\hat{k} \in C^\infty$ and $\kappa \in C^\infty([0,1] \times S^{2n})$, so we transfer the seminorms of this Frechet space to obtain a complementary set of seminorms on Ψ^m.

It is simple consequence of proposition 2.6 that

$$K_j \in OP\Psi^{m_j} \Rightarrow K_1K_2 \in OP\Psi^{m_1+m_2} \quad \text{and} \quad K_1^* \in OP\Psi^{m_1}. \tag{2.60}$$

Note the remark following the proof of proposition 2.6. We also easily obtain the following result.

Theorem 2.17. Let $K \in OP\Psi^m$. Suppose K_0, in the expansion (2.56), satisfies the conditions that $\sigma_{K_0}(\pm 1)(X,D)$ are elliptic and both are injective (resp. surjective, resp. invertible) on $\mathcal{S}(\mathbb{R}^n)$. Then K has a left (resp. right, resp. two sided) parametrix $L \in OP\Psi^{-m}$.

Proof. Take the case $\sigma_{K_0}(\pm 1)(X,D)$ injective. Then theorem 2.10 gives $L_0 \in OP\Psi_0^{-m}$ such that $L_0K_0 = I$. It follows that

$$L_0K = I + S_1, \; S_1 \in OP\Psi^{-1}. \tag{2.61}$$

Then we can construct $L \in OP\Psi^{-m}$, such that

$$L \sim (I - S_1 + S_1^2 - \cdots)L_0, \tag{2.62}$$

and then $LK = I$ modulo a smoothing operator. The other assertions of this theorem follow similarly.

For example, $\mathcal{L}_0+i\alpha T + P$, with $P \in OP\Psi^1$, has a parametrix on $OP\Psi^{-2}$, provided α satisfies the condition (2.46). In the case of scalar differential operators, the regularity result contained in theorem 2.17 was obtained by Miller [M10], via energy estimates, rather than a symbolic construction of a parametrix.

Unlike the situation in proposition 2.12, if $K \in OP\Psi^m$ as in theorem 2.17, with $\sigma_{K_0}(\pm 1)(X,D)$ elliptic but not both injective, it is still possible for K to be hypoelliptic. We consider the following example, due to Stein [S7]; see also Rothschild [R4]. Let

$$K = \mathcal{L}_0+i\alpha T+\beta \in OP\Psi^2 \tag{2.63}$$

where α is chosen to be an integer such that the condition (2.46) for hypoellipticity of $K_0 = \mathcal{L}_0+i\alpha T$ is violated, $\beta \in \mathbb{C}$. Let $S \in OP\Psi_o^{0,\infty}$ be the projection produced by the proof of proposition 2.12. Then, as in (2.34), we have $E_o \in OP\Psi_o^{-2}$ such that

$$E_o(\mathcal{L}_0+i\alpha T) = I - S, \tag{2.64}$$

while

$$S(\mathcal{L}_0+i\alpha T) = 0. \tag{2.65}$$

Then, if $\beta \neq 0$,

$$\begin{aligned}(E_o + \beta^{-1}S)K &= (E_o + \beta^{-1}S)(K_0 + \beta) \\ &= I + \beta E_0.\end{aligned} \tag{2.66}$$

Note that $R = \beta E_0 \in OP\Psi_o^{-2,\infty}$. It follows that

$$E \sim (I-R+R^2-\cdots)(E_o + \beta^{-1}S) \in OP\Psi^0 \tag{2.67}$$

gives a left parametrix of K, so the operator (2.63) is hypoelliptic, with loss of two derivatives, whenever $\beta \neq 0$. A generalization, in the case of right invariant differential operators on $\mathbb{H}^n$, was given by Rothschild, and will be discussed in section 4 of this chapter. We will look at further generalizations of this phenomenon in Chapter III.

The machinery summarized in theorem 2.17 and proposition 2.14 is particularly effective in examining hypoellipticity and constructing parametrices for a second order differential operator P (right invariant) on $\mathbb{H}^n$, which is doubly characteristic on the set $\Lambda \in T^*\mathbb{H}^n$ of characteristics of $\mathcal{L}_0$. Note that Λ is characterized as a line bundle which is translation invariant and its fiber over the identity $0 \in \mathbb{H}^n$ is the linear span of dt. Suppose the principal symbol of P is positive and vanishes to exactly second order on Λ. It follows that the principal symbol of P must agree with that of

$$\sum_{j,k=1}^{2n} a_{jk} X_j X_k = P_2, \tag{2.68}$$

where we have set

$$X_j = L_j,\ X_{j+n} = M_j,\ 1 \le j \le n. \tag{2.69}$$

The matrix (a_{jk}) is symmetric and positive definite. Since $L_j, M_j \in OP\Psi^1_0$, by (2.5), it follows that

$$P = \sum_{j,k} a_{jk} X_j X_k + i\alpha T + B, \quad B \in OP\Psi^1, \tag{2.70}$$

for some $\alpha \in \mathbb{C}$. The sum of the first two terms belongs to $OP\Psi^2_0$, and we have

$$\sigma_{P_2 + i\alpha T}(\pm 1)(X,D) = -\Sigma a_{jk} \chi_j \chi_k \mp \alpha \tag{2.71}$$

where we have set

$$\chi_j = x_j,\ \chi_{j+n} = \xi_j,\ 1 \le j \le n. \tag{2.72}$$

In particular, the ellipticity hypothesis of theorem 2.17 holds, and proposition 2.14 applies in a straightforward fashion to the question of invertibility. Non-translation-invariant generalizations form a very important class of operators, which will be investigated in Chapter III.

In order to achieve a genuine microlocal analysis, it is important to amalgamate the operator classes $OP\Psi^m$ and their non-right-invariant extensions, with the classical operator classes OPS^m. We begin by considering right invariant operators in OPS^m from the point of view of the representations of $\mathbb{H}^n$. We shall discuss connections with propositions 1.12-1.18 of Chapter I in section 1 of Chapter III.

<u>Definition 2.18.</u> We say $\hat{k} \in \Sigma_0^m$ provided $\hat{k}(\tau,y,\eta)$ is C^∞ away from $(\tau,y,\eta) = 0$, and

$$\hat{k}(\lambda\tau,\lambda y,\lambda\eta) = \lambda^m k(\tau,y,\eta), \quad \lambda > 0. \tag{2.73}$$

In such a case, we say $Ku = k*u$ defines $K \in OP\Sigma_0^m$. If $\hat{k} \sim \sum_{j\geq 0} \hat{k}_j$, $\hat{k}_j \in \Sigma_0^{m-j}$, we say $\hat{k} \in \Sigma^m$ and $K \in OP\Sigma^m$.

The proof of proposition 1.4 from Chapter I immediately implies that, modulo smoothing operators, $OP\Sigma^m$ is identical with the class of right invariant operators in OPS^m. Note that (2.73) is equivalent to

$$\sigma_K(\underline{+}\lambda)(x,\xi) = \lambda^m\sigma_K(\underline{+}1)(\lambda^{-\frac{1}{2}}x,\lambda^{-\frac{1}{2}}\xi). \tag{2.74}$$

Parallel to proposition 2.2, we have the following simple result, whose proof we omit.

<u>Proposition 2.19.</u> The formula (2.64) defines an element of Σ_0^m if and only if $\sigma_K(\underline{+}1)(x,\xi)$ are smooth with asymptotic behavior

$$\sigma_K(\underline{+}1)(x,\xi) \sim \sum_{j\geq 0} (\underline{+}1)^j\varphi_{m-j}(\underline{+}x,\xi), \tag{2.75}$$

where $\varphi_{m-j}(x,\xi)$ is homogeneous of degree m-j in (x,ξ).

Of course, the operator calculus for OPS^m implies that if $P_j \in OP\Sigma^{m_j}$, then $P_1P_2 \in OP\Sigma^{m_1+m_2}$. We want to perceive this via the representation theory of $\mathbb{H}^n$. The argument proving proposition 2.6, on products of elements of $OP\Psi^{m_j}$, will have to be modified, since, in (2.74), the argument of

$\sigma_K(\pm 1)$ is $(\lambda^{-\frac{1}{2}}x,\lambda^{-\frac{1}{2}}\xi)$. In fact, in this case, letting $\lambda \to 0$ causes technical problems we wish to avoid. One way to avoid this problem is to restrict attention to operators defined by (2.74) with $\sigma_K(\pm 1)(x,\xi) \in C_o^\infty(\mathbb{R}^{2n})$. It is clear that this subset of OPS^m contains operators that agree with any right invariant operator in OPS^m, microlocally on a conic neighborhood of the line bundle Λ (characteristic set of $\mathcal{L}_0$) discussed above. That this is a sufficiently rich class for our purposes is guaranteed by the following result.

Proposition 2.20. If $K \in OP\Psi^m$, then, microlocally on the complement of any conic neighborhood of $\Lambda \subset T^*\mathbb{H}^n \setminus 0$, K belongs to OPS^m.

Proof. Given the formula (1.23) from chapter I, analyzing K as a pseudo-differential operator, we see this is a simple consequence of the following observation. If $\hat{k}(\tau,y,\eta) \in \Psi_o^m$, then, on the complement of any conic neighborhood of $\{y=\eta=0\}$ in $\mathbb{R}^{2n+1} \setminus 0$, $\hat{k}$ agrees with an element of Σ^m.

Let us now consider the symbolic calculus applied to a product of $A \in OP\Sigma_o^m$, $B \in OP\Sigma_o^\mu$, where

$$\text{(2.76)}\quad \sigma_A(\pm 1)(x,\xi) = a_\pm(x,\xi) \text{ and } \sigma_B(\pm 1)(x,\xi) = b_\pm(x,\xi) \text{ belong to } C_o^\infty(\mathbb{R}^{2n}).$$

Let us denote

$$\text{(2.77)}\qquad a_{\pm,\lambda}(x,\xi) = a_\pm(\lambda^{-\frac{1}{2}}x,\lambda^{-\frac{1}{2}}\xi),$$

with $b_{\pm,\lambda}(x,\xi)$ similarly defined. Consequently,

$$\text{(2.78)}\qquad \sigma_{AB}(\pm\lambda)(X,D) = \lambda^{m+\mu}\, e_\pm(\lambda,X,D),$$

with

$$\text{(2.79)}\qquad e_\pm(\lambda,x,\xi) = \sum_{j=0}^{N} (1/j!)\{a_{\pm,\lambda},b_{\pm,\lambda}\}_j(x,\xi) + \lambda^{-N}R_N^\pm(\lambda,x,\xi).$$

Recall that $\{a,b\}_j$ is given by (2.20). To see that we can avoid considering $\lambda \to 0$, note that, if A' is defined by

$$(2.80)\qquad \sigma_{A'}(\underline{+}\lambda)(X,D) = \psi(\lambda)\sigma_A(\underline{+}\lambda)(X,D)$$

where $\psi(\lambda)$ is smooth, $\psi(\lambda) = 1$ for $|\lambda| \geq 1$, 0 for $|\lambda| \leq \frac{1}{2}$, then A and A' differ by an operation of convolution by w, where $\hat{w}(\tau,y,\eta)$ has compact support. Hence $w \in C^\infty(\mathbb{H}^n)$, so A and A' differ by a smoothing operator. Hence, in (2.78), (2.79), we can restrict attention to $|\lambda| \geq 1$. We can rewrite (2.79) as

$$(2.81)\qquad e_{\underline{+}}(\lambda,X,D) = \sum_{j=0}^{N} \lambda^{-j}\, e_j^{\underline{+}}(\lambda^{-\frac{1}{2}}X,\lambda^{-\frac{1}{2}}D) + \lambda^{-N}R_N^{\underline{+}}(\lambda,X,D).$$

Results of [H10] imply that $R_N^{\underline{+}}(\lambda,X,D)$ is bounded in $OP\mathcal{S}_1^0$ (defined by (2.16)), for $|\lambda| \geq 1$. To see that the remainder in (2.81) makes a contribution which is smoothing to a high degree if N is large, consider the following. $T \in OP\Sigma_o^0$ has symbol

$$(2.82)\qquad \sigma_T(\underline{+}\lambda)(X,D) = \underline{+}i\lambda.$$

Thus $A(BT^k)$ has a form similar to (2.78), (2.81), with $\lambda^{m+\mu}$ replaced by $(\underline{+}i)^k\lambda^{m+\mu+k}$; in fact,

$$(2.83)\qquad \sigma_{ABT^k}(\underline{+}\lambda)(X,D) = \lambda^{m+\mu}\sum_{j=0}^{N} \lambda^{-j} e_j^{\pm}(\lambda^{-\frac{1}{2}}X,\lambda^{-\frac{1}{2}}D)T^k + (\underline{+}i)^k\lambda^{m+\mu+k-N}\, R_N^{\pm}(\lambda,X,D).$$

COnsequently (2.78),(2.81) give AB as a sum

$$(2.84)\qquad AB = \sum_{j=1}^{N} E_j + S_N$$

where $E_j \in OP\Sigma_o^{m+\mu-j}$ and S_N has the following property:

$$(2.85)\qquad T^{k_1}S_NT^{k_2}: L^2(\mathbb{H}^n) \to L^2(\mathbb{H}^n) \quad \text{for } N \geq m+\mu+k_1+k_2.$$

Since S_N is certainly a classical pseudodifferential operator in $OPS^{m+\mu}$, whose symbol is essentially supported near Λ, and since $T \in OPS^1$ is mocrolocally elliptic near Λ, this implies that S_N is arbitrarily smoothing for N large. Consequently (2.84) is an asymptotic relation, so, for $A \in OP\Sigma_o^m$, $B \in OP\Sigma_o^\mu$, we have $AB \in OP\Sigma^{m+\mu}$, and

(2.86) $$\sigma_{AB}(\pm\lambda)(X,D) \sim \lambda^{m+\mu} \sum_{j\geq 0} \lambda^{-j}\, e_j^{\pm}(\lambda^{-\frac{1}{2}}X,\lambda^{-\frac{1}{2}}D).$$

Note that

$$e_0^{\pm}(x,\xi) = a_{\pm}(x,\xi)b_{\pm}(x,\xi),$$

which verifies that (for scalar operators)

(2.87) $$[A,B] = AB - BA \in OP\Sigma^{m+\mu-1}.$$

Of course, examining products of operators in $OP\Sigma^{m_j}$ via the Heisenberg group harmonic analysis involves a bit more work than just appealing to the OPS^m operator calculus. The point is that it does work to produce a symbol calculus for such operators, whereas commutative harmonic analysis does not work to produce a symbol calculus for $OP\Psi^m$. The calculations just done should be viewed as a warm-up for what is to come.

Now let us consider the products PK and KP, given $K \in OP\Psi_o^{\mu}$, $P \in OP\Sigma_o^m$, the symbol of P being supported near Λ. Thus

(2.88) $$\sigma_{PK}(\pm\lambda)(X,D) = \lambda^{\mu/2+m}\, p_{\pm}(\lambda^{-\frac{1}{2}}X,\lambda^{-\frac{1}{2}}D)a_{\pm}(X,D).$$

As above, we can restrict attention to $|\lambda| \geq 1$. As in the analysis of (2.78), we have

(2.89) $$\sigma_{PK}(\pm\lambda)(X,D) = \lambda^{\mu/2+m}\, e_{\pm}(\lambda,X,D)$$

where, with $p_{\pm,\lambda}(x,\xi) = p_{\pm}(\lambda^{-\frac{1}{2}}x,\lambda^{\frac{1}{2}}\xi)$,

(2.90) $$e_{\pm}(\lambda,x,\xi) = \sum_{j=0}^{N} (1/j!)\{p_{\pm,\lambda},a_{\pm}\}_j(x,\xi) + \lambda^{-N/2}\, R_N^{\pm}(\lambda,x,\xi).$$

As before, $R_N^{\pm}(\lambda,x,\xi)$ is bounded in $\mathcal{S}_1^{\mu}$ for $\lambda \geq 1$.

In this case, the formula (2.20) for $\{a,b\}_j$ gives

(2.91) $$e_{\pm}(\lambda,x,\xi) = \sum_{j=1}^{N}\sum_{k=1}^{K(j)} p_{kj}^{\pm}(\lambda^{-\frac{1}{2}}x,\lambda^{-\frac{1}{2}}\xi)a_{kj}(x,\xi)\lambda^{-j/2} + \lambda^{-N/2}\, R_N^{\pm}(\lambda,x,\xi),$$

where

$$(2.92)\qquad p^{\pm}_{kj}(x,\xi) \in C^{\infty}_{o}(\mathbb{R}^n), \quad a^{\pm}_{kj}(x,\xi) \in \mathcal{H}^{\mu-j}_{b}.$$

The same argument as before shows the remainder in (2.91) contributes a term smoothing to high order. To summarize, we have:

Proposition 2.21. If $K \in OP\Psi^{\mu}_{o}$, $P \in OP\Sigma^{m}_{o}$, the symbol of P being supported near Λ, then

$$(2.93)\qquad \sigma_{PK}(\underline{+}\lambda)(X,D) \sim \lambda^{\mu/2+m} \sum_{j\geq 0} \lambda^{-j/2}\, e^{\pm}_{j}(\lambda,X,D)$$

where

$$(2.94)\qquad e^{\pm}_{0}(\lambda,x,\xi) = p_{\pm}(\lambda^{-\frac{1}{2}}x,\lambda^{-\frac{1}{2}}\ )a_{\pm}(x,\xi)$$

and generally

$$(2.95)\qquad e^{\pm}_{j}(\lambda,x,\xi) = \sum_{k=1}^{K(j)} p^{\pm}_{kj}(\lambda^{-\frac{1}{2}}x,\lambda^{-\frac{1}{2}}\xi)a^{\pm}_{kj}(x,\xi),$$

with (2.92) holding. In a similar fashion we have

$$(2.96)\qquad \sigma_{KP}(\underline{+}\lambda)(X,D) \sim \lambda^{\mu/2+m} \sum_{j\geq 0} \lambda^{-j/2}\, f^{\pm}_{j}(\lambda,X,D)$$

where

$$(2.97)\qquad f^{\pm}_{0} = e^{\pm}_{0}$$

and $f^{\pm}_{j}(\lambda,x,\xi)$ has an expression similar in form to (2.95).

Note that this analysis gives for the commutation [P,K]:

$$(2.98)\qquad \sigma_{[P,K]}(\underline{+}\lambda)(X,D) \sim \lambda^{m+\mu/2} \sum_{j\geq 1} \lambda^{-j/2}[e_{j}(\lambda,X,D)-f_{j}(\lambda,X,D)].$$

Inductively, from the representation (2.93), (2.94), we obtain:

Corollary 2.22. If $K \in OP\Psi^{\mu}_{o}$, $P \in OP\Sigma^{m}_{o}$ are as above, then

$$(2.99)\qquad [P,K] \sim \sum_{j\geq 1} \sum_{k=1}^{K(j)} P_{kj}A_{kj}, \quad P_{kj} \in OP\Sigma^{m}, \ A_{kj} \in OP\Psi^{\mu-j}.$$

It is worth noting that more precise statements can be made about PK when $K \in OP\Psi^{\mu,\infty}$. Of course, elements of $OP\Psi^{\mu,\infty}$ have special properties, one of which is the following.

Proposition 2.23. If $K \in OP\Psi_o^{\mu,\infty}$, then, microlocally on the complement of any conic neighborhood of Λ, K belongs to $OPS^{-\infty}$.

Proof. In view of the analysis of K as a pseudodifferential operator given in Chapter I, proposition 1.1, this follows from the observation that, if $\hat{k}(\tau,y,\eta) \in \Psi_o^{\mu,\infty}$, then $\hat{k}(\tau,y,\eta)$ is rapidly decreasing outside any conic neighborhood of $\{y=\eta=0\}$ in $\mathbb{R}^{2n+1}$.

Now if $P \in OP\Sigma^m$ is as above and $K \in OP\Psi_o^{\mu,\infty}$, the formulas (2.88)-(2.92) hold, and in addition we have

$$a_{kj}^{\pm}(x,\xi) \in \mathcal{S}(\mathbb{R}^{2n}). \tag{2.100}$$

We can expand $p_{kj}(\lambda^{-\frac{1}{2}}x,\lambda^{-\frac{1}{2}}\xi)$ in a power series, in powers of $\lambda^{-\frac{1}{2}}$. Indeed,

$$p_{kj}^{\pm}(\lambda^{-\frac{1}{2}}x,\lambda^{-\frac{1}{2}}\xi) = \sum_{\ell=0}^{N} (1/\ell!)\lambda^{-\ell/2}\, \Omega^{\ell} p_{kj}^{\pm}(0,0) + \lambda^{-N/2} r_{kjN}^{\pm}(\lambda,x,\xi) \tag{2.101}$$

where

$$\Omega p(y,\eta) = (x,\xi)\cdot\nabla_{y,\eta} p(y,\eta) \tag{2.102}$$

and

$$r_{kjN}^{\pm}(\lambda,x,\xi) \text{ is bounded in } \mathcal{S}_1^N \text{ for } \lambda \geq 1. \tag{2.103}$$

Since the factors $a_{kj}^{\pm}(x,\xi)$ in (2.92) belong to $\mathcal{S}(\mathbb{R}^n) = \mathcal{S}_1^{-\infty}$ in this case, we have

$$\sigma_{PK}(\underline{+}\lambda)(X,D) = \lambda^{m+\mu/2}\, e_{\pm}(\lambda,X,D) \tag{2.104}$$

with

$$e_{\pm}(\lambda,X,D) \sim \sum_{j\geq 0} \lambda^{-j/2}\, e_j^{\pm}(X,D), \quad e_j^{\pm}(x,\xi) \in \mathcal{S}(\mathbb{R}^{2n}). \tag{2.105}$$

Note that

$$e_0^{\pm}(x,\xi) = p_{\pm}(0,0)a_{\pm}(x,\xi). \tag{2.106}$$

We have proved the following result.

Proposition 2.25. If $P \in OP\Sigma^m$ and $K \in OP\Psi_o^{\mu,\infty}$, then $PK \in OP\Psi^{2m+\mu,\infty}$. Moreover, modulo a smoothing operator, $PK = L$, with

$$\sigma_L(\pm\lambda)(x,\xi) \sim \sum_{j\geq 0} \lambda^{m+\mu/2-j/2} e_j^{\pm}(x,\xi), \tag{2.108}$$

where $e_j^{\pm}(x,\xi) \in \mathcal{S}(\mathbb{R}^{2n})$ and

$$e_0^{\pm}(x,\xi) = p_{\pm}^0 \sigma_K(\pm 1)(x,\xi). \tag{2.109}$$

The factor $p_{\pm}^0$ is equal to the principal symbol of P, evaluated at the point $\pm dt$ lying in the fiber of $T^*\mathbb{H}^n$ over the origin. The product $KP \in OP\Psi^{2m+\mu,\infty}$ has a similar behavior.

Note that, in light of proposition 2.23, we need not make the assumption that the symbol of P is essentially supported near Λ. From proposition 2.25 we obtain the following simple corollary.

Corollary 2.26. If $K \in OP\Psi^{\mu,\infty}$, then there exists $A \in OP\Sigma^{\mu/2}$ such that

$$K = AK_0 \tag{2.110}$$

with $K_0 \in OP\Psi^{0,\infty}$.

It follows that elements of $OP\Psi^{\mu,\infty}$ enjoy better continuity properties on Sobolev spaces than general elements of $OP\Psi^{\mu}$, if $\mu > 0$.

We now consider a certain synthesis of the classical (right invariant) pseudodifferential operator classes $OP\Sigma^*$ and the operator classes $OP\Psi^*$. We will first consider a class of operators which are microlocally supported in a small conic neighborhood of the line bundle Λ in $T^*\mathbb{H}^n \setminus 0$. In light of our calculations involving one parameter families of pseudodifferential operators on $\mathbb{R}^n$, the following class is quite natural.

Definition 2.27. We say A belongs to $OP\Omega_b^{m,k}$ provided $b_{\pm}(\lambda,x,\xi) = \sigma_A(\pm\lambda)(x,\xi)$ satisfies the following two conditions:

(2.111) $b_{\pm}(\lambda,x,\xi)$ is smooth and supported on $|x|+|\xi| \leq C|\lambda|^{\frac{1}{2}}$, $|\lambda| \geq C'$,

and

$$\lambda^{j-m} D_\lambda^j b_\pm(\lambda,\cdot,\cdot) \text{ is bounded in } \mathcal{S}_1^k \text{ for } \lambda \in \mathbb{R}^+. \tag{2.112}$$

Recall that the symbol class $\mathcal{S}_1^k$ is defined by (2.16). Hence (2.112) is equivalent to

$$|D_\lambda^j D_x^\beta D_\xi^\alpha b_\pm(\lambda,x,\xi)| \le C_{j\alpha\beta} \lambda^{m-j}(1+|x|+|\xi|)^{k-|\alpha|-|\beta|}. \tag{2.113}$$

If Au = a*u, then, according to (1.16), we have

$$\hat{a}(\pm\tau,y,\eta) = b_\pm(\tau,\pm\tau^{-\frac{1}{2}}y,\tau^{-\frac{1}{2}}\eta). \tag{2.114}$$

Hence we see that the support condition in (2.114) is equivalent to:

$$\hat{a}(\tau,y,\eta) \text{ is supported on } |y|+|\eta| \le C|\tau|,\ |\tau| \ge C'. \tag{2.115}$$

As for the content of (2.112), we have the following.

<u>Proposition 2.28</u>. The hypothesis (2.112) is equivalent to the following estimate on a:

$$\begin{aligned} |D_\tau^j D_y^\beta D_\eta^\alpha \hat{a}(\tau,y,\eta)| &\le C_{j\alpha\beta}\tau^{m-j}(\tau^{\frac{1}{2}}+|y|+|\eta|)^{k-|\alpha|-|\beta|}\tau^{-k/2} \\ &= C_{j\alpha\beta}\tau^{m-j-\frac{1}{2}(|\alpha|+|\beta|)}(1+|\tau^{-\frac{1}{2}}y|+|\tau^{-\frac{1}{2}}\eta|)^{k-|\alpha|-|\beta|}. \end{aligned} \tag{2.116}$$

<u>Proof</u>. It is straightforward to differentiate the right side of (2.114) with respect to y and η. We get

$$\begin{aligned} |D_y^\beta D_\eta^\alpha \hat{a}(\tau,y,\eta)| &= \tau^{-\frac{1}{2}(|\alpha|+|\beta|)}|D_x^\beta D_\xi^\alpha b_\pm(\tau,\pm\tau^{-\frac{1}{2}}y,\tau^{-\frac{1}{2}}\eta)| \\ &\le C\tau^{-\frac{1}{2}(|\alpha|+|\beta|)}\tau^m(1+|\tau^{-\frac{1}{2}}y|+|\tau^{-\frac{1}{2}}\eta|)^{k-|\alpha|-|\beta|}, \end{aligned} \tag{2.117}$$

thereby verifying (2.116) in case j = 0. Since it is not so easy to apply D_τ^j to the right side of (2.114), we proceed more carefully. Let us say that $b_\pm(\lambda,x,\xi)$ belongs to $\Omega(m,k)$ if (2.111) and (2.112) hold. Now applying one τ derivative to (2.114) gives

$$\begin{aligned} D_\tau b_\pm(\tau,\pm\tau^{-\frac12}y,\tau^{-\frac12}\eta) &= D_\lambda b_\pm(\tau,\pm\tau^{-\frac12}y,\tau^{-\frac12}\eta) \\ &\mp \tfrac12\tau^{-3/2}\, yD_x b_\pm(\tau,\pm\tau^{-\frac12}y,\tau^{-\frac12}\eta) \\ &- \tfrac12\tau^{-3/2}\eta D_\xi b_\pm(\tau,\pm\tau^{-\frac12}y,\tau^{-\frac12}\eta). \end{aligned}$$

Now, note that $c_\pm(\lambda,x,\xi) \in \Omega(m,k)$ implies

$$D_\lambda c_\pm,\ \lambda^{-1}\ xD_x c_\pm,\ \text{and}\ \lambda^{-1}\xi D_\xi c_\pm\ \text{belong to}\ \Omega(m-1,k).$$

We deduce that, if $b_\pm(\lambda,x,\xi) \in \Omega(m,k)$, then

$$D_\tau b_\pm(\tau,\pm\tau^{-\frac12}y,\tau^{-\frac12}\eta) = \sum_\ell C_{\ell\pm}(\tau,\pm\tau^{-\frac12}y,\tau^{-\frac12}\eta)$$

with $C_{\ell\pm}(\lambda,x,\xi) \in \Omega(m-1,k)$. It follows by induction that

$$D_\tau^j b_\pm(\tau,\pm\tau^{-\frac12}y,\tau^{-\frac12}\eta) = \sum_\ell C_{j\ell\pm}(\tau,\pm\tau^{-\frac12}y,\tau^{-\frac12}\eta)$$

with

$$C_{j\ell\pm}(\lambda,x,\xi) \in \Omega(m-j,k).$$

In view of this, the estimate (2.117) shows that (2.116) follows from (2.112). The converse result is proved by a similar argument.

The estimates (2.116) show that

$$\hat{a}(\tau,y,\eta) \in S^{m+\kappa/2}_{\frac12}, \qquad \kappa = \max(k,0), \tag{2.118}$$

and, outside any conic neighborhood of $\{y=\eta=0\}$, a satisifes the estimates for membership in $S^{m+k/2}_{1\#}$. Recalling the support condition (2.115), we deduce the following from proposition 1.1 and (1.23) of Chapter I.

<u>Proposition 2.29.</u> If $A \in OP\Omega_b^{m,k}$, then

$$A \in OPS^{m+\kappa/2}_{\frac12,\frac12}, \qquad \kappa = \max(k,0). \tag{2.119}$$

Furthermore, A belongs to $OPS^{m+k/2}_{1,0}$ microlocally outside and conic neighborhood of the line bundle $\Lambda \subset T^*\mathbb{H}^n \setminus 0$, and, outside a certain conic neighborhood $\mathfrak{V}$ of Λ, A is microlocally in $OPS^{-\infty}$.

In view of this, it is reasonable to introduce the following classes of operators.

Definition 2.30. We say a right invariant operator A belongs to $OP\Omega^{m,k}$ provided that A coincides with an element of $OP\Omega_b^{m,k}$ microlocally on some conic neighborhood of Λ and A belongs to $OPS_{1,0}^{m+k/2}$ outside any conic neighborhood of Λ.

The following result on the symbol calculus for $OP\Omega_b^{m,k}$ follows easily from the properties of $OP\mathcal{S}_1^*$.

Proposition 2.31. If $A \in OP\Omega_b^{m,k}$, $B \in OP\Omega_b^{m',k'}$, then, modulo a smoothing operator, $AB \in OP\Omega_b^{m+m',k+k'}$. Also, $A^* \in OP\Omega_b^{m,k}$.

Proof. We have $AB \in OPS_{\frac{1}{2},\frac{1}{2}}^{m+m'+\frac{1}{2}(\kappa+\kappa')}$, and

$$\sigma_{AB}(\underline{+}\lambda)(X,D) = \sigma_A(\underline{+}\lambda)(X,D)\sigma_B(\underline{+}\lambda)(X,D). \tag{2.120}$$

It is clear that $\sigma_{AB}(\underline{+}\lambda)(x,\xi)$ satisfies the condition (2.112), with m replaced by m+m', k replaced by k+k'. The support condition (2.111) is not quite verified, but since both A and B are microlocally in $OPS^{-\infty}$ outside $\mathfrak{v}$, by proposition 2.29, so is their product, so one can subtract a smoothing operator from AB to recover the support property (2.111). The proof for A^* is immediate.

A straightforward amalgamation of the symbol calculus (2.120) and the usual operator calculus for $OPS_{1,0}^m$ handles adjoints and products of operators in $OP\Omega^{m,k}$. We should point out that, in view of the characterization (2.116), $OP\Omega^{m,k}$ is contained in Boutet de Monvel's class

$$OPS^{m+k/2,k}(\mathbb{H}^n,\Lambda). \tag{2.121}$$

See Boutet de Monvel [B7] for a development of these operator classes. In fact, $OP\Omega^{m,k}$ coincides (modulo smoothing operators) with the right invariant operators in Boutet de Monvel's class. However, when the machinery of Chapter I is implemented to produce $OP\tilde{\Omega}^{m,k}$ (see Chapter III), this class will be a strict subclass of $OPS^{m+k/2,k}(\mathbb{H}^n,\Lambda)$, and will have the advantage of possessing the symbol calculus we have developed.

Let us note the following inclusions of operator classes (which are quite evident):

(2.122) $$OP\Psi^m \subset OP\Omega^{m/2,m},$$

(2.123) $$OP\Psi^{m,k} \subset OP\Omega^{m/2,m-k} \qquad (k \geq 0)$$

(2.124) $$OP\Sigma^m \subset OP\Omega^{m,\mu}, \qquad \mu = \max(0,m).$$

The classes $OP\Omega^{m,k}$ have associated hypoellipticity results, extending theorem 2.17. Since this section has already grown fairly long, we will postpone discussing such results until Chapter III.

3. Functional calculus for the Heisenberg Laplacian and for the harmonic oscillator Hamiltonian.

In section 2 we considered inverses of the operators

(3.1) $$\mathcal{L}_\alpha = \mathcal{L}_0 + i\alpha T$$

where

(3.2) $$\mathcal{L}_0 = \Sigma(L_j^2+M_j^2)$$

is the "Heisenberg Laplacian." We saw that $\mathcal{L}_\alpha$ is hypoelliptic, with inverse in $OP\Psi_o^{-2}$, provided α avoids the set

(3.3) $$\{\ldots,-n-2,-n-1,-n\} \cup \{n,n+1,n+2,\ldots\}.$$

Here we would like to understand the behavior of more general functions $f(\mathcal{L}_\alpha)$. There is some overlap between the material of this section and the material presented in Chapter I, section 7, of [T5], but here we will concentrate more on the technical aspects having to do with the theory of pseudodifferential operators. Recall that

(3.4) $$\sigma_{\mathcal{L}_\alpha}(\pm\lambda)(X,D) = -\lambda(-\Delta+|x|^2\pm\alpha).$$

It follows that

$$\sigma_{f(\mathcal{L}_\alpha)}(\pm\lambda)(X,D) = f(-\lambda(-\Delta+|x|^2\pm\alpha)). \tag{3.5}$$

Thus we need to understand

$$f(H), \quad H = -\Delta + |x|^2 \tag{3.6}$$

for a general class of functions of the harmonic oscillator Hamiltonian $H = -\Delta + |x|^2$.

We begin by computing the Weyl symbol of the operator simigroup

$$e^{-tH} = h_t(X,D). \tag{3.7}$$

Formula (3.8) below was given by Peetre [P1]; see also Unterberger [U1].

Proposition 3.1. We have

$$\tilde{h}_t(q,p) = c_n'(\sinh t)^{-n} \exp[-(|q|^2+|p|^2)\coth t] \tag{3.8}$$

and

$$h_t(x,\xi) = (\cosh t)^{-n} \exp[-(|x|^2+|\xi|^2) \tanh t]. \tag{3.9}$$

Proof. As a first step, note that, by commutativity,

$$e^{-tH} = e^{-tH_1} \cdots e^{-tH_n}$$

where

$$H_j = -\partial^2/\partial x_j^2 + x_j^2.$$

Thus $h_t(x,\xi) = h_t(x_1,\xi_1)\cdots h_t(x_n,\xi_n)$, and $\tilde{h}_t(q,p)$ satisfies the analogous multiplicativity condition. Since the right sides of (3.8) and (3.9) are also multiplicative, it suffices to prove the proposition for $H = -d^2/dx^2 + x^2$, acting on functions of one variable. Now the Weyl symbol $h_t(x,\xi)$ is related to the kernel of the operator e^{-tH}, defined by

$$e^{-tH} u(x) = \int K_t(x,y)u(y)dy, \tag{3.10}$$

by

$$K_t(x,y) = \int e^{i(x-y)\xi} h_t(\tfrac{1}{2}(x+y),\xi)d\xi. \tag{3.11}$$

Consequently, the identity (3.9) (for n=1) is equivalent to

$$K_t(x,y) = (2\pi)^{-\frac{1}{2}}(\sinh 2t)^{-\frac{1}{2}} \exp\{[-\tfrac{1}{2}(\coth 2t)(x^2+y^2)-xy]/\sinh 2t\}, \tag{3.12}$$

which in turn is equivalent to the well known generating function identity

$$\sum_{j=0}^{\infty} h_j(x)h_j(y)t^j = \pi^{-\frac{1}{2}}(1-t^2)^{-\frac{1}{2}} \exp\{[2xyt+(x^2+y^2)t^2]/(1-t^2)\}e^{-x^2/2-y^2/2} \tag{3.13}$$

for Hermite functions (see Lebedev [L4], pp. 61-63). This proves (3.9), and (3.8) follows by taking the Fourier transform.

A different proof of proposition 3.1, closer to that of Peetre, utilizing the Bargmann-Fok representation of $\mathbb{H}^n$, is given in Chapter I of [T5]; still another derivation is given in Appendix A of this paper; see (A.18).

Now we can make an analytic continuation of (3.7)-(3.9) to Re t > 0, and pass to the limit as t becomes purely imaginary. In that way we get formulas for the Weyl symbol of e^{itH}, $t \in \mathbb{R}$ We have

$$e^{itH} = E_t(X,D) \tag{3.14}$$

with

$$E_t(x,\xi) = (\cos t)^{-n} \exp[-i(\tan t)(|x|^2+|\xi|^2)], \tag{3.15}$$

at least when t is different from a half integral multiple of π. At such singular values, $E_t(x,\xi)$ achieves a limiting value in $\mathcal{S}'(\mathbb{R}^{2n})$.

One simple consequence of (3.15) is the following analysis of a general class of operators as pseudodifferential operators.

<u>Proposition 3.2</u>. Let $f(\lambda) \in S^m_{\rho\,\#}(\mathbb{R})$, i.e., suppose $|f^{(j)}(\lambda)| \leq C_j(1+|\lambda|)^{m-\rho j}$. If $\frac{1}{2} \leq \rho \leq 1$, then

$$f(H) = F(X,D) \in OP\mathcal{S}^{2m}_{\rho'}, \qquad \rho' = 2\rho-1. \tag{3.16}$$

i.e., $|D^\beta_x D^\alpha_\xi F(x,\xi)| \leq C_{\alpha\beta}(1+|x|+|\xi|)^{2m-\rho'(|\alpha|+|\beta|)}$, and in fact

$$(3.17)\qquad F(x,\xi) = f(|x|^2+|\xi|^2) + r(|x|^2+|\xi|^2)$$

with

$$(3.18)\qquad r(|x|^2+|\xi|^2) \in \mathcal{S}_{\rho'}^{2m-2(2\rho-1)}$$

Proof. If $f(\lambda) \in S^m_{\rho\#}(R)$, we can write $f = f_1 + f_2$ where $\hat{f}_1 \in \mathcal{E}'(R)$ is supported on $|t| < \pi/4$ and $f_2 \in \mathcal{S}(R)$, the Schwartz space of rapidly decreasing functions. Hence $f(H) = f_1(H) + f_2(H)$. Now

$$(3.19)\qquad \begin{aligned} f_2(H) &= \int \hat{f}_2(t)e^{itH}dt \\ &= (-iH)^{-k} \int \hat{f}_2^{(k)}(t)e^{itH}dt. \end{aligned}$$

It follows that $f_2(H)$ is a smoothing operator, and

$$(3.20)\qquad \|H^k f_2(H)\| \le \int |f_2^{(k)}(t)|dt.$$

As for $f_1(H)$, we have

$$(3.21)\qquad f_1(H) = \int \hat{f}_1(t)e^{itH}dt,$$

and, using (3.15) with $|t| < \pi/4$, we have $f_1(H) = F_1(X,D)$, with

$$(3.22)\qquad F_1(x,\xi) = 2^{-n} \int \hat{f}_1(\arctan s)\ (1+s^2)^{n/2}\ e^{is(|x|^2+|\xi|^2)}\ ds.$$

Here the distribution $\hat{f}_1(\arctan s)$ is written formally. Thus

$$(3.23)\qquad F_1(x,\xi) = \varphi(|x|^2+|\xi|^2)$$

with

$$(3.24)\qquad \varphi(\lambda) = 2^{-n} \int \hat{f}_1(\arctan s)\ (1+s^2)^{n/2}\ e^{is\lambda}\ ds.$$

We see this is equal to the symbol (at s=0) of a pseudodifferential operator obtained by applying a change of variable to $f_1(D_s) \in OP^m_{\rho,0}(R)$. Thus, if $\rho \ge \frac{1}{2}$,

$$(3.25)\qquad \varphi(\lambda) - f(\lambda) \in S^{m-(2\rho-1)}_{\rho\#}.$$

Our conclusions (3.17) and (3.18) follow immediately from this.

If $\rho > \frac{1}{2}$, a complete asymptotic expansion can be given for $\varphi(\lambda)$. Note in particular that, if $f(\lambda)$ has an asymptotic expansion

$$(3.26) \qquad f(\lambda) \sim \sum_{j\geq 0} f_{m-j}(\lambda),$$

with f_{m-j} homogeneous of degree m-j, then

$$(3.27) \qquad F(x,\xi) \sim f(|x|^2+|\xi|^2) + \sum_{j\geq 1} r_{2m-2j}(x,\xi) \in \mathcal{H}_b^{2m}.$$

It is well known that, if $F(x,\xi) \in S^0_{1,0}(\mathbb{R}^n)$, and in particular if $F(x,\xi) \in \mathcal{S}^0_1$, then $F(X,D)$ is continuous on $L^p(\mathbb{R}^n)$, for $1 < p < \infty$. See, e.g., Nagase [N2] or Stein [S8]. As a consequence we deduce that, if $f(\lambda) \in S^0_{1\#}$, then $f(H)\colon L^p(\mathbb{R}^n) \to L^p(\mathbb{R}^n)$, for $1 < p < \infty$. This result was proved by Mauceri [M2] by different means, involving setting up a Littlewood-Paley theory based on the semigroup e^{-tH}.

Proposition 3.2 is a special case of a result proved by Helffer and Robert [H4], by different means. The method of proof here, via

$$(3.28) \qquad f(H) = \int \hat{f}(t)\, e^{itH}\, dt$$

is a modification of the method used by the author to derive a functional calculus for first order elliptic self adjoint pseudodifferential operators; see Chapter XII of [T2], and also [T3]. Compare some of the arguments of Appendix B of this paper.

One class of operators which proposition 3.2 helps us analyze is the class of fractional powers of $-\mathcal{L}_0$. Note that, for $\alpha \in \mathbb{R}$,

$$(3.29) \qquad \sigma_{(-\mathcal{L}_0)^{\alpha}}(\underline{+}\lambda)(X,D) = \lambda^{\alpha}(-\Delta+|x|^2)^{\alpha} = \lambda^{\alpha}H^{\alpha}.$$

Now proposition 3.2, together with (3.27), gives

$$(3.30) \qquad (-\Delta+|x|^2)^{\alpha} \in OP\mathcal{H}_b^{2\alpha},$$

if we let $f(\lambda)$ be an even element of $S^{\alpha}_{1\#}$ equal to $|\lambda|^{\alpha}$ for $|\lambda| \geq \frac{1}{2}$ and smoothed out for $|\lambda| \leq \frac{1}{2}$; since $|\lambda| \leq \frac{1}{2}$ does not intersect the spectrum of H, $H^\alpha = f(H)$. Now, we claim

(3.31) $$H^\alpha \sim \sum_{j\geq 0} \varphi_{2j}(|x|^2+|\xi|^2)$$

where $\varphi_0(\lambda) = \lambda^\alpha$ and $\varphi_{2j}(\lambda)$ is homogeneous of degree $\alpha-2j$ in λ; in other words, the odd terms in (3.26) are absent.

To prove (3.31) we may as well suppose $\alpha < 0$, since (3.31) could then be deduced for general α by multiplication by integral powers of H. In that case, we have $H^\alpha = F_1(X,D)$ mod $OP\mathcal{S}_1^{-\infty}$, with $F_1(X,D)$ given by (3.22) and

(3.32) $$\hat{f}_1(\arctan s) = \psi(s)\,|\arctan s|^{1-\alpha}$$
$$= \psi(s)\,|s|^{1-\alpha}g(s^2)$$

where $\psi \in C_0^\infty(-\frac{1}{4}\pi,\frac{1}{4}\pi)$, $\psi(s) = 1$ for $|s| \leq \pi/8$, say, and $g(\lambda)$ is analytic. Since the factors $(1+s^2)^{n/2}$ and $g(s^2)$ are both analytic in s^2, for $|s| \leq \pi/4$, the result (3.31) follows. This gives

(3.33) $$\sigma_{(-\mathcal{L}_0)^\alpha}(\underline{+}1)(X,D) = H^\alpha \in OP\mathcal{H}^{2\alpha},$$

and hence proposition 2.2 applies, to give:

<u>Proposition 3.3.</u> We have

(3.34) $$(-\mathcal{L}_0)^\alpha \in OP\Psi_0^{2\alpha}.$$

For more complicated functions of $\mathcal{L}_0$, proposition 3.2 is not such an incisive tool, since it is not so effective in studying $f(\lambda H)$ for a large parameter λ, unless λ factors out, as it does in (3.29). However, using the identities given in proposition 3.1, we can analyze the solution operator

(3.35) $$e^{s\mathcal{L}_0}$$

to the "Heisenberg group heat equation"

$$(3.36)\qquad \partial u/\partial s = \mathcal{L}_0 u.$$

In fact, we have

$$(3.37)\qquad e^{s\mathcal{L}_0}\delta_0(t,q,p) = k_s(t,q,p)$$

with

$$(3.38)\qquad \hat{k}_s(\pm\tau,y,\eta) = \sigma_{e^{s\mathcal{L}_0}}(\pm\tau)(\pm\tau^{-\frac{1}{2}}y,\tau^{-\frac{1}{2}}\eta)$$

$$= h_{s\tau}(\pm\tau^{-\frac{1}{2}}y,\tau^{-\frac{1}{2}}\eta),$$

where $h_t(x,\xi)$ is given by (3.9). Hence, if

$$(3.39)\qquad (\mathcal{F}_1 k_s)(\lambda,q,p) = \int_{-\infty}^{\infty} e^{-it\lambda} k_s(t,q,p)\,dt,$$

we have, by (3.8), $(\mathcal{F}_1 k_s)(\tau,q,p) = c_n\tau^n\ (\sinh s\tau)^{-n}\ \exp[-(\tau\coth\tau)(|q|^2+|p|^2)]$, so

$$(3.40)\quad k_s(t,q,p) = c_n s^{-n-1}\int_{-\infty}^{\infty} e^{i\tau(t/s)}(\tau/\sinh\tau)^n \exp[-(\tau\coth\tau)(|q|^2+|p|^2)/s]\,d\tau$$

$$= s^{-n-1}\ k_1(t/s,q/\sqrt{s},p/\sqrt{s})$$

where

$$(3.41)\quad k_1(t,q,p) = c_n\int_{-\infty}^{\infty} e^{it\tau}(\tau/\sinh\tau)^n\ \exp[-(\tau\coth\tau)(|q|^2+|p|^2)]\,d\tau.$$

Let us state this formally.

Proposition 3.4. For $s > 0$, we have

$$(3.42)\qquad e^{s\mathcal{L}_0}\delta_0(t,q,p) = k_s(t,q,p)$$

with

$$(3.43)\qquad k_s(t,q,p) = s^{-n-1}\ k_1(t/s,q/\sqrt{s},p/\sqrt{s})$$

and $k_1 \in \mathcal{S}(\mathbb{R}^n)$ given by (3.41).

This result was obtained by B. Gaveau [G1] by a different method, utilizing a diffusion process construction. It was also obtained, from a representation theory point of view, by Hulanicki [H14] and in the Ph.D.

thesis of Geller [G4]. From (3.40)-(3.43) it is a simple matter to calculat the kernel

(3.44) $$(-\mathcal{L}_0)^{-\frac{1}{2}} e^{-s(-\mathcal{L}_0)^{\frac{1}{2}}} \delta_0(t,q,p) = P_s(t,q,p)$$

using the subordination identity

(3.45) $$y^{-1} e^{-xy} = \pi^{\frac{1}{2}} \int_0^\infty e^{-x^2/4v} e^{-vy^2} v^{-\frac{1}{2}} dv.$$

We obtain

(3.46) $$P_s(t,q,p) = c_n' \int_{-\infty}^{\infty} (\tau/\sinh \tau)^n [\tfrac{1}{4}s^2 + (\tau \coth \tau)(|q|^2+|p|^2) - it\tau]^{-n-\frac{1}{2}} d\tau.$$

We have developed certain aspects of the functional calculus for $\mathcal{L}_0$. It is useful to treat more general operators, of the form

(3.47) $$P = \Sigma a_{jk} X_j X_k + i\alpha T = P_o + i\alpha T,$$

with (a_{jk}) positive definite, whose hypoellipticity was discussed in (2.68)-(2.72) of section 2, as a consequence of proposition 2.14. To construct the heat kernel, we can proceed as in (3.37)-(3.41), provided we know the Weyl symbol $h_t^Q(x,\xi)$ of

(3.48) $$e^{-tQ(X,D)} = h_t^Q(X,D).$$

Here $Q(x,\xi) = \Sigma a_{jk} \chi_j \chi_k$ as in (2.72). Now, if $Q(x, \ )$ takes the form

(3.49) $$Q(x,\xi) = \Sigma \mu_j (x_j^2 + \xi_j^2) \qquad (\mu_j > 0),$$

which is equivalent to having

(3.50 $$P_o = \Sigma \mu_j (L_j^2 + M_j^2),$$

then the proof of proposition 3.1 gives

(3.51) $$h_t^Q(x,\xi) = \prod_{j=1}^n (\cosh t\,\mu_j)^{-1} \exp -[\sum_{j=1}^n (x_j^2+\xi_j^2)(\tanh t\,\mu_j)] .$$

Consequently for P_o of the form (3.50), we have

(3.52) $$e^{sP_Q}\delta_0(t,q,p) = k_s^Q(t,q,p)$$

with

(3.53) $$k_s^Q(t,q,p) = s^{-n-1}\, k_1^Q(t/s,q/\sqrt{s},p/\sqrt{s}),$$

where $k_1^Q \in \mathcal{S}(\mathbb{R}^{2n+1})$ is given by

(3.54) $$k_1^Q(t,q,p) = c_n \int_{-\infty}^{\infty} e^{it\tau} \prod_{j=1}^{n} (\tau/\sinh \mu_j\tau)\exp[-\sum_{j=1}^{n} (\tau \coth \mu_j\tau)(q_j^2+p_j^2)]d\tau.$$

Now, given any P_o of the form $\Sigma a_{jk}X_jX_k$, we can pick a symplectic basis of $\mathbb{R}^{2n}$ diagonalizing the quadratic form $Q(x,\xi) = \Sigma a_{jk}\chi_j\chi_k$. The resulting automorphism of $\mathbb{H}^n$ puts P_o in the form (3.49), so the formula (3.54) for the heat kernel is valid generally, upon applying an automorphism to $\mathbb{H}^n$. If $a_{jk}(y)$ depends smoothly on a parameter y, automorphisms of $\mathbb{H}^n$ putting $P_o(y)$ into diagonal form may not be chosen to depend smoothly on y, so it is desirable to express the kernel (3.54) in invariant form. We do this using the Hamilton map F_Q associated with the quadratic form Q, as defined in section 2, i.e.,

(3.55) $$Q(u,v) = i^{-1}\, \sigma(u,F_Q v),$$

where Q(u,v) is the symmetric bilinear form such that Q(u,u) = Q(u). The eigenvalues of F_Q are of the form $\pm\mu_j$, where μ_j are as in (3.49), so

(3.56) $$\det \sinh \tau F_Q = -(\prod_{j=1}^{n} \sinh \mu_j\tau)^2,$$

and hence

(3.57) $$\prod_{j=1}^{n} (\tau/\sinh \mu_j\tau) = (-\tau^{-2n} \det \sinh \tau F_Q)^{-\frac{1}{2}}.$$

Now let

(3.58) $$A_Q = (F_Q^2)^{\frac{1}{2}},$$

the unique square root of F_Q^2 with positive spectrum. Then, with

(3.59) $\quad (q,p) = z,$

we can write

(3.60) $$-\tau \sum_{j=1}^{n} (\coth \mu_j\tau)(q_j^2+p_j^2) = -\tau Q(A_Q^{-1} \coth \tau A_Q z,z).$$

Consequently we can write the heat kernel (3.54) invariantly as

(3.61) $$k_1^Q(t,z) = c_n \int_{-\infty}^{\infty} e^{it\tau}\, \Phi_Q(\tau,z)\, d\tau$$

with

(3.62) $$\Phi_Q(\tau,z) = (-\tau^{-2n} \det \sinh \tau F_Q)^{-\frac{1}{2}} \exp[-\tau Q(A_Q^{-1} \coth \tau A_Q z,z)].$$

Note that, if $P_\alpha = P_o + i\alpha T$,

(3.63) $$\begin{aligned} e^{sP_\alpha}\delta_0(t,p,q) &= e^{sP_o}\delta_0(t+is\alpha,q,p) \\ &= s^{-n-1}\, k_1^Q(\tfrac{t}{s} + i\alpha, q/\sqrt{s}, p/\sqrt{s}) \end{aligned}$$

where $k_1^Q(t+i\alpha,q,p)$ is defined from (3.61) by analytic continuation as long as

(3.64) $$|\mathrm{Re}\ \alpha| < \sum_{j=1}^{n} \mu_j = \tfrac{1}{2} \operatorname{tr} A_Q.$$

Let us return to $\mathcal{L}_0$, and analyze the wave equation by analytically continuing (3.46). First, with $z = q+ip \in \mathbb{C}^n$, we can rewrite (3.46) as

(3.65) $$P_s(t,z) = c_n \int_\gamma (\tau/\sin \tau)^n [s^2 + g_{A,B}(\tau)]^{-n-\frac{1}{2}}\, d\tau$$

where the path γ is the imaginary axis, from $-i\infty$ to $+i\infty$, and

(3.66) $$g_{A,B}(\zeta) = B\,\zeta \cot \zeta + A\zeta,$$

with

((3.67) $\quad A = 4t,\ B = 4|z|^2.$

We can analytically continue $P_s(t,z)$ to $\mathrm{Re}\ s > 0$, and pass to purely imaginary s, by deforming the contour γ. If γ is deformed to γ' in such a way that its image under $g_{A,B}$ hugs part of the positive real axis, then singularities in the deformed path γ' (where $g'_{A,B} = 0$) will correspond to singularities in the solution operator

$$(-\mathcal{L}_0)^{-\frac{1}{2}}\, e^{is(-\mathcal{L}_0)^{\frac{1}{2}}}$$

to the "Heisenberg group wave equation." We refer to Chapter I, section 8, of [T5] for details on this, but record the result here. Such a result had been obtained by Nachman [N1], by a different method.

Proposition 3.5. The fundamental solution of the wave equation

$$(3.68)\qquad (-\mathcal{L}_0)^{-\frac{1}{2}}\, e^{is(-\mathcal{L}_0)^{\frac{1}{2}}}\delta_0(t,z)$$

has singularities only where, for some j,

$$(3.69)\qquad s^2 = g_{A,B}(x_j(A,B))$$

where $g_{A,B}(\zeta)$ is given by (3.66), (3.67), and the points $x_j(A,B)$ are the real zeros of $g'_{A,B}(x)$.

Such a result is also a special case of results on propagation of singularities proved by Melrose [M6] and Lascar [L1] for $\partial^2/\partial t^2 - P$ when P has symplectic characteristics of codimension two (including (3.68) on the three dimensional Heisenberg group) and Lascar and Lascar [L2] in greater generality. These results imply the group $e^{is(-\mathcal{L}_0)^{\frac{1}{2}}}$ moves the wave front set of a distribution by the unique continuous (not smooth) extension of the Hamilton flow defined by the symbol of $(-\mathcal{L}_0)^{\frac{1}{2}}$ on the complement of the characteristic set. In particular, if P is any pseudodifferential operator on $\mathbb{H}^n$, the conjugate $e^{-is(-\mathcal{L}_0)^{\frac{1}{2}}}P\, e^{is(-\mathcal{L}_0)^{\frac{1}{2}}} = P(s)$ must preserve wave front sets; $WF(P(s)u) \subset WF(u)$ for any $u \in \mathcal{E}'(\mathbb{H}^n)$. It is consequently of interest to analyze the conjugated operator as some sort of pseudodifferential operator, thus generalizing Egorov's theorem, which would describe such a conjugated operator if $\mathcal{L}_0$ were elliptic. We will look at some examples here where P is a convolution operator on $\mathbb{H}^n$. We will see that such conjugated operators are outside the classes of pseudodifferential operators on $\mathbb{H}^n$ we have defined so far, and further classes of operators will arise.

To start with what seems to be the simplest example, suppose $P \in OP\Psi_0^{m,\infty}$, so

$$\sigma_P(\pm\lambda)(X,D) = \lambda^m A_\pm, \qquad A_\pm \in OP\mathcal{S}_1^{-\infty}. \tag{3.70}$$

It follows that

$$P(s) = e^{-is(-\mathcal{L}_0)^{\frac{1}{2}}}\, P\, e^{is(-\mathcal{L}_0)^{\frac{1}{2}}} \tag{3.71}$$

satisfies

$$\begin{aligned} \sigma_{P(x)}(\pm\lambda)(X,D) &= \lambda^m\, e^{-is\lambda^{\frac{1}{2}}H^{\frac{1}{2}}}\, A_\pm e^{is\lambda^{\frac{1}{2}}H^{\frac{1}{2}}} \\ &= \lambda^m\, B_\pm(s,\lambda). \end{aligned} \tag{3.72}$$

Note that, for any integer k, the operator norm of $H^k\ B_\pm(s,\lambda)H^k$ on $L^2(\mathbb{R}^n)$ is bounded independently of λ, so, for $0 < \lambda < \infty$, $B_\pm(s,\lambda)$ is bounded in $OP\mathcal{S}_1^{-\infty}$ (as s runs over a bounded interval). If we take λ derivatives, we see that

$$D_\lambda^j B(s,\lambda) = \lambda^{-j/2}\, e^{-is\lambda^{\frac{1}{2}}H^{\frac{1}{2}}}[(-\text{ad}\ H^{\frac{1}{2}})^j (A_\pm)]e^{is\lambda^{\frac{1}{2}}H^{\frac{1}{2}}} + \cdots \tag{3.73}$$

where the finite number of lower order terms in (3.73) all involve lower order powers of λ. Hence, for general $A_\pm$, $\lambda^{j/2} e^{is\lambda^{\frac{1}{2}}H^{\frac{1}{2}}}\ D_\lambda^j\ B(s,\lambda)e^{-is\lambda^{\frac{1}{2}}H^{\frac{1}{2}}}$ tends to a nonzero limit as $\lambda \to \infty$, so the best that can be said is:

$$\lambda^{j/2}\ D_\lambda^j\ B_\pm(s,\lambda) \quad \text{is bounded in } OP\mathcal{S}_1^{-\infty}, \text{ for } 1 \le \lambda < \infty. \tag{3.74}$$

Note that cutting off (3.72) on some bounded λ-interval alters P(s) by a smoothing operator in this case, so we need only worry about $|\lambda| \ge 1$. Note that (3.74) is weaker than the hypothesis (2.112), in case $k = -\infty$, so P(s) does not belong to $OP\Omega^{m,-\infty}$.

This example suggests the following class of operators. We say A belongs to $OP\Omega_{\frac{1}{2}b}^{m,k}$ provided $b_\pm(\lambda,x,\xi) = \sigma_A(\pm\lambda)(x,\xi)$ satisfies the conditions

(3.75) $b_{\pm}(\lambda, x, \xi)$ is smooth and supported on $|x| + |\xi| \le C|\lambda|^{\frac{1}{2}}$, $|\lambda| \ge C'$,

and

(3.76) $\lambda^{\frac{1}{2}j-m} D_{\lambda}^{j} b_{\pm}(\lambda,\cdot,\cdot)$ is bounded in $\mathcal{S}_{1}^{k}$, for $\lambda \in \mathbb{R}^{+}$.

Note the parallel with the conditions (2.111)-(2.112), defining $OP\Omega_{b}^{m,k}$. The analysis proving proposition 2.28 shows that, for $Au = a*u$, the hypothesis (3.76) implies

$$(3.77) \quad |D_{\tau}^{j} D_{y}^{\beta} D_{\eta}^{\alpha} a(\tau,y,\eta)| \le C_{j\alpha\beta} \tau^{m-\frac{1}{2}j-\frac{1}{2}(|\alpha|+|\beta|)} (1+|\tau^{-\frac{1}{2}}y|+|\tau^{-\frac{1}{2}}\eta|)^{k-|\alpha|-|\beta|}.$$

Thus we see that

$$(3.78) \quad OP\Omega_{b}^{m,k} \subset OP\Omega_{\frac{1}{2}\,b}^{m,k} \subset OPS_{\frac{1}{2},\frac{1}{2}}^{m+\kappa/2}. \qquad \kappa = \max\,(k,0).$$

Also, in view of (3.77), it is reasonable to say a right invariant operator A belongs to $OP\Omega_{\frac{1}{2}}^{m,k}$ if it can be written as a sum of an element of $OP\Omega_{\frac{1}{2}\,b}^{m,k}$ and an element of $OPS_{1,0}^{m+\frac{1}{2}k}$. It follows that

$$(3.79) \quad OP\Omega^{m,k} \subset OP\Omega_{\frac{1}{2}}^{m,k} \subset OPS_{\frac{1}{2},\frac{1}{2}}^{m+\kappa/2}, \qquad \kappa = \max(k,0).$$

We should remark that $OP\Omega_{\frac{1}{2}}^{m,k}$ is not contained in Boutet de Monvel's class $OPS^{m+\frac{1}{2}k,k}(\mathbb{H}^n,\Lambda)$. If we let $OP\Omega_{\frac{1}{2}}^{m,-\infty} = \bigcap_{k>-\infty} OP\Omega_{\frac{1}{2}}^{m,k}$, then what we have seen is that

$$(3.80) \quad P \in OP\Psi^{m,\infty} \Rightarrow P(s) \in OP\Omega_{\frac{1}{2}}^{m,-\infty},$$

where P(s) is given by (3.71). More generally, one sees without much trouble that

$$(3.81) \quad P \in OP\Omega_{\frac{1}{2}}^{m,-\infty} \Rightarrow P(s) \in OP\Omega_{\frac{1}{2}}^{m,-\infty}.$$

The assertions (3.80) and (3.81) follow directly from (3.73) and its natural generalization with $A_{\pm}$ replaced by $A_{\pm}(\lambda)$, in view of the fact that

$$(3.82) \quad E \in OP\mathcal{S}_{1}^{-\infty} \Rightarrow e^{-itH^{\frac{1}{2}}}\, E\, e^{itH^{\frac{1}{2}}} \text{ is bounded in } OP\mathcal{S}_{1}^{-\infty},\ t \in \mathbb{R}^{+}.$$

That this is true follows from the fact that a complete set of seminorms defining the topology of $OPS_1^{-\infty}$ is given by the sequence of L^2-operator norms $P_k(E) = \|H^k E H^k\|$, $k = 0,1,2,\ldots$, each of which is invariant under conjugation by $e^{itH^{\frac{1}{2}}}$.

For finite k, it does not seem that $P \in OP\Omega_{\frac{1}{2}}^{m,k}$ implies $P(s) \in OP\Omega_{\frac{1}{2}}^{m,k}$. To see what sort of operator P(s) is, we need to understand

$$E(t) = e^{itH^{\frac{1}{2}}} E_0 e^{-itH^{\frac{1}{2}}}, \qquad t = s\lambda^{\frac{1}{2}}, \tag{3.83}$$

given $E_0 \in OPS_1^k$. We need to have an analysis as $|t| \to \infty$ as well as for finite t. A good analysis for $|t|$ bounded is provided as follows. Set

$$E_0(t) = E_0, \tag{3.84}$$

and, for $j \geq 1$, let $E_j(t)$ be defined by

$$dE_j/dt = i[H^{\frac{1}{2}}, E_{j-1}(t)]. \tag{3.85}$$

Then, by induction, we see that

$$E_j(t) = E_j t^j = (i^j/j!)(\mathrm{ad}\ H^{\frac{1}{2}})^j E_0\ t^j. \tag{3.86}$$

Note that

$$E_j \in OPS_1^{k-j}. \tag{3.87}$$

If we write

$$E(t) = \sum_{j=0}^{N} E_j t^j + R_N(t) = F_N(t) + R_N(t) \tag{3.88}$$

we see that

$$dF_N/dt = i[H^{\frac{1}{2}}, F_N(t)] - it^N[H^{\frac{1}{2}}, E_N], \tag{3.89}$$

from which it follows, by Duhamel's principle, that

$$R_N(t) = (i^{N+1}/N!) \int_0^t e^{isH^{\frac{1}{2}}} [(\mathrm{ad}\ H^{\frac{1}{2}})^{N+1} \cdot E_0] e^{-isH^{\frac{1}{2}}} [(t-s)^N] ds. \tag{3.90}$$

Let us introduce the following notions. Set

(3.91) $\mathcal{D}_s = H^{-s/2}(L^2(\mathbb{R}^n))$,

and say

(3.92) $T \in \mathcal{O}(m) \Leftrightarrow T: \mathcal{D}_s \to \mathcal{D}_{s-m}$ for all $s \in \mathbb{R}$.

Note that

(3.93) $OP\mathcal{S}_1^m \subset \mathcal{O}(m)$,

so we have $E_j \in \mathcal{O}(k-j)$ if $E_0 \in OP\mathcal{S}_1^k$, and, for each N,

(3.94) $(1+|t|)^{-N-1}R_N(t)$ is bounded in $\mathcal{O}(k-N)$, for $t \in \mathbb{R}$.

This is not a very incisive result, due to the blow-up of $R_N(t)$ as $|t| \to \infty$.

We can do a little better by using a geometrical optics approach. Note that, if $e(t,x,\xi)$ is the full symbol of $E(t)$, it satisfies

(3.95) $$\partial e/\partial t \sim H_g e + \sum_{\substack{j\geq 3\\ \text{odd}}} (2i/j!)\{g,e\}_j$$

where $\{g,e\}_j$ is given by (2.20) and g is the symbol of $H^{\frac{1}{2}}$:

(3.96) $g(X,D) = H^{\frac{1}{2}}$.

We have $g \in \mathcal{H}_b^1$; in fact, by (3.31),

(3.97) $g(x,\xi) \sim g_1(x,\xi) + g_2(x,\xi) + \cdots$

with $g_{1+j}(x,\xi)$ homogeneous of degree 1-4j and

(3.98) $g_1(x,\xi) = (|x|^2+|\xi|^2)^{\frac{1}{2}}$.

We produce

(3.99) $$e(t,x,\xi) \sim \sum_{j\geq 0} e_j(t,x,\xi)$$

with $e_0(t,x,\xi)$ satisfying

(3.100) $\partial e_0/\partial t = H_{g_1} e_0$; $e(0,x,\xi) = E_0(x,\xi)$, so

$$(3.101)\qquad e_0(t,x,\xi) = E_0(\chi(t)(x,\xi))$$

where

$$(3.102)\qquad \chi(t) = \exp tH_{g_1} = \exp(t/g_1)\Sigma\partial/\partial\theta_j;$$

here

$$(3.103)\qquad \Sigma\partial/\partial\theta_j = \tfrac{1}{2} H_{|x|^2+|\xi|^2}$$

generates a group

$$(3.104)\qquad \rho(t) = \exp(t\ \Sigma\partial/\partial\theta_j)$$

of rotations on $\mathbb{R}^{2n}$, and

$$(3.105)\qquad e_0(t,x,\xi) = E_0(\rho(t/g_1)(x,\xi)).$$

Note that $e_0(t,x,\xi)$ satisfies estimates of the form

$$(3.106)\qquad |D_\xi^\alpha D_x^\beta e_0(t,x,\xi)| \le C_{\alpha\beta}(1+|x|+|\xi|)^{k-|\alpha|-|\beta|}(1+|t|/g_1)^{|\alpha|+|\beta|}.$$

Then

$$(3.107)\qquad (\partial e_0/\partial t)(t,X,D) - i[H^{\frac{1}{2}},e_0(t,X,D)] = A_1(t,X,D)$$

where $A_1(t,x,\xi)$ satisfies estimates of the form

$$(3.108)\qquad |D_\xi^\alpha D_x^\beta A_1(t,x,\xi)| \le C_{\alpha\beta}(1+|x|+|\xi|)^{k-5-|\alpha|-|\beta|}(1+|t|/g_1)^{3+|\alpha|+|\beta|}.$$

One can hence solve for $e_1(t,x,\xi)$ the equation

$$(3.109)\qquad \partial e_1/\partial t - H_{g_1} e_1 = -A_1(t,x,\xi),\ e_1(0,x,\xi) = 0,$$

and continue in this fashion, to get a more accurate approximation to $e(t,X,D) = E(t)$. Such an approach still does not give an incisive analysis of (3.71) for $P \in OP\Psi_o^{m,k}$, with k finite, but it does allow one to analyze (3.71) in case $P \in OP\Sigma^m$ has symbol vanishing on a conic neighborhood of Λ. In that case, we are reduced to understanding (3.83) when the symbol of E_0 is supported on

$$M^{-1}\lambda^{\frac{1}{2}} \leq (|x|^2+|\xi|^2)^{\frac{1}{2}} \leq M\lambda^{\frac{1}{2}},$$

and, as the right sides of (3.106) and (3.108) indicate, the growth in t does not present a problem in this case. Since the symbol of P is supported where $\mathcal{L}_0$ is elliptic, in this case, this analysis merely reproduces standard results of geometrical optics, so we will not dwell on it.

The correct class of (not necessarily right invariant) pseudodifferential operators, invariant under conjugation by such unitary operators as $e^{is(-\mathcal{L}_0)^{\frac{1}{2}}}$, and the corresponding extension of Egorov's theorem, remain to be achieved.

4. A hypoellipticity result of Rothschild.

As we saw in section 2, if $K_0 \in OP\Psi^m_o$ has the property that $\sigma_{K_0}(\pm 1)(X,D)$ are elliptic but not both injective, K_0 will not be hypoelliptic, but it is possible for $K = K_0 + K_1$, $K_1 \in OP\Psi^{m-1}$, to be hypoelliptic. In [R4], Rothschild proved the following surprising result, giving a complete analysis of hypoellipticity for such operators that are differential operators.

Proposition 4.1. Let K be a right invariant differential operator on $\mathbb{H}^n$. Suppose $K = K_0 + K_1$ with $K_0 \in OP\Psi^m_o$, $K_1 \in OP\Psi^{m-1}$, and $\sigma_{K_0}(\pm 1)(X,D)$ elliptic. Suppose furthermore that, for some M,

(4.1) $\sigma_K(\pm\lambda)(X,D)$ is injective, whenever $|\lambda| \geq M$.

Then K is hypoelliptic.

This section will be devoted to a proof of the following natural generalization of proposition 4.1.

Proposition 4.2. Suppose $K \in OP\Psi^m$ is a pseudodifferential operator given by a finite sum

(4.2) $K = K_0 + K_1 + \cdots + K_\mu,\ K_j \in OP\Psi_o^{m-j}.$

Suppose $\sigma_{K_0}(\pm 1)(X,D)$ elliptic, and suppose (4.1) holds. Then K is hypoelliptic.

Our proof of proposition 4.2 will be along the same lines as Rothschild's proof, with some simplifications, due to the use of the machinery developed in section 2.

Of course, the hypoellipticity of K fails to follow from Theorem 2.17 only in the case when one of the operators $\sigma_{K_0}(\pm 1)(X,D)$ has a nontrivial kernel. Let $\pi_\pm$ denote the orthogonal projections onto these kernels (one of these operators may be zero, but not both).

We make some preliminary simplifications. First, we can assume K is self adjoint, since the hypoellipticity of K^*K implies the hypoellipticity of K, and hypothesis (4.1) for K implies the same sort of hypothesis for K^*K. Thus each K_j in (4.2) is self adjoint, and $\sigma_{K_j}(\pm\lambda)(X,D)$ is formally self adjoint. Now, if we define $\tilde{K}_0 \in OP\Psi_o^m$ by

(4.3) $\sigma_{\tilde{K}_0}(\pm\lambda)(X,D) = \sigma_{K_0}(\pm\lambda)(X,D) + \lambda^{m/2}\pi_\pm$

then $\tilde{K}_0$ has an inverse $\tilde{G}_0 \in OP\Psi_o^{-m}$, and we obtain a new operator

(4.4) $$\begin{aligned} E = \tilde{G}_0 K &= \tilde{G}_0 K_0 + \cdots + \tilde{G}_0 K \\ &= E_0 + E_1 + \cdots + E_\mu \in OP\Psi^0, \end{aligned}$$

with $E_j \in OP\Psi_o^{-j}$. To prove hypoellipticity of K, it will suffice to construct a parametrix for E. Note that

(4.5) $\sigma_{E_0}(\pm 1)(X,D) = I - \pi_\pm.$

By the observation above, to construct a parametrix for E, it suffices to do so for E^*E, so we can assume without loss of generality that

(4.6) $E \in OP\Psi^0$ is self adjoint, with expansion (4.4), and the operator we have still satisfies (4.5) and, in addition:

(4.7) $\sigma_E(\pm\lambda)(X,D)$ is invertible, for $|\lambda| \geq M$.

Consequently, we are reduced to constructing a parametrix for E, under assumptions (4.5)-(4.7). Note that $\pi_{\pm}$ are projections onto finite dimensional subspaces of $\mathcal{S}(\mathbb{R}^n)$ and hence are operators in $OP\mathcal{S}_1^{-\infty}$; this follows from general results on the kernel of an elliptic operator in $OP\mathcal{H}_b^0 \subset OP\mathcal{S}_1^0$; see Grusin [G12] or Beals [B4].

Note that, if

(4.8) $\sigma_{E_j}(\pm 1)(X,D) = A_j^{\pm} \in OP\mathcal{H}^{-j} \subset OP\mathcal{S}_1^{-j}$,

then

(4.9) $\sigma_E(\pm\lambda)(X,D) = I - \pi_{\pm} + \sum_{j=1}^{\mu} \lambda^{-j/2} A_j^{\pm}$.

The operators $A_j^{\pm}$ in (4.9) are all compact and self adjoint. Now, if we set

(4.10) $\varepsilon = \lambda^{-\frac{1}{2}}, \ |\varepsilon| \leq M^{\frac{1}{2}}$,

then, according to (4.9), $\sigma_E(\pm\lambda)(X,D)$ is equal to

(4.11) $P_{\pm}(\varepsilon) = I - \pi_{\pm} + \sum_{j=1}^{\mu} A_j^{\pm}\varepsilon^j = I - \pi_{\pm} + A^{\pm}(\varepsilon)$,

an analytic function (in fact, a polynomial) in ε, taking values in $OP\mathcal{H}^0$. Our hypothesis (4.7) states that the operators $P_{\pm}(\varepsilon)$ are invertible on $L^2(\mathbb{R}^n)$, for sufficiently small ε. We wish to study these inverses.

Clearly, $A_{\pm}(\varepsilon)$ has small operator norm if $|\varepsilon|$ is small, so the spectrum of $P_{\pm}(\varepsilon)$ is concentrated near the points 0 and 1. In particular, if γ denotes the circle of radius $\frac{1}{2}$ centered about 0, then for $|\varepsilon|$ small no points of the spectrum of $P_{\pm}(\varepsilon)$ lie within a distance $\frac{1}{4}$ of γ. Let us set

(4.12) $\tilde{\pi}_{\pm}(\varepsilon) = (2\pi i)^{-1} \int_{\gamma} (\zeta - P_{\pm}(\varepsilon))^{-1} \, d\zeta$,

and

(4.13) $$Q_{\pm}(\varepsilon) = I - \tilde{\pi}_{\pm}(\varepsilon)$$

$$= (2\pi i)^{-1} \int_{\gamma'} (\zeta - P_{\pm}(\varepsilon))^{-1}\, d\zeta,$$

where γ' denotes the circle of radius $\frac{1}{2}$ centered at $\zeta = 1$. The operators $\pi_{\pm}(\varepsilon)$ are analytic in ε, taking values in $\mathcal{L}(L^2(\mathbb{R}))$, and are all projections. Note that $\tilde{\pi}_{\pm}(\varepsilon) \to \pi_{\pm}$ in operator norm as $\varepsilon \to 0$, so $\tilde{\pi}_{\pm}(\varepsilon)$ are all projections onto finite dimensional spaces, of dimension equal to the dimension of the range of $\pi_{\pm}$, for ε small. Let us write

(4.14) $$\tilde{\pi}_{\pm}(\varepsilon) = \sum_{j=0}^{\infty} \kappa_j^{\pm} \varepsilon^j \qquad (\kappa_0^{\pm} = \pi_{\pm}),$$

so $\kappa_j^{\pm}$ are bounded operators on $L^2(\mathbb{R}^n)$.

<u>Lemma 4.3</u>. $\kappa_j^{\pm} \in OP\mathcal{S}_1^{-\infty}$.

<u>Proof</u>. The operators $Q_{\pm}(\varepsilon)$ are analytic in ε with values in $OP\mathcal{H}^0$, all with principal symbol equal to 1, so in fact

$$Q_{\pm}(\varepsilon) = I + B_{\pm}(\varepsilon)$$

with $B_{\pm}(\varepsilon)$ analytic in ε, with values in $OP\mathcal{S}_1^{-1}$. Hence, for each N, we have an analytic family $F_N(\varepsilon)$ in $OP\mathcal{S}_1^0$ such that

$$F_N^{\pm}(\varepsilon) Q_{\pm}(\varepsilon) = I + R_N^{\pm}(\varepsilon)$$

where $R_N^{\pm}(\varepsilon)$ is analytic in ε with values in $OP\mathcal{S}_1^{-N}$. In fact, take

$$F_N(\varepsilon) = I - B_{\pm}(\varepsilon) + \cdots + (-B_{\pm}(\varepsilon))^N.$$

Now we have

$$0 = F_N(\varepsilon) Q_{\pm}(\varepsilon) \tilde{\pi}_{\pm}(\varepsilon)$$

$$= (I + R_N^{\pm}(\varepsilon)) \tilde{\pi}_{\pm}(\varepsilon),$$

so

$$\tilde{\pi}_{\pm}(\varepsilon) = - R_N^{\pm}(\varepsilon)\tilde{\pi}_{\pm}(\varepsilon).$$

It follows that the operators $\kappa_j^{\pm}$ in (4.14) belong to $\bigcap_N \mathcal{S}_1^{-N} = \mathcal{S}_1^{-\infty}$, so the proof is complete.

Suppose that $W_{\pm}(\varepsilon)$ inverts $P_{\pm}(\varepsilon)$ on the range of $\tilde{\pi}_{\pm}(\varepsilon)$, and

$$W_{\pm}(\varepsilon) = \tilde{\pi}_{\pm}(\varepsilon)W_{\pm}(\varepsilon) = W_{\pm}(\varepsilon)\tilde{\pi}_{\pm}(\varepsilon). \tag{4.15}$$

This uniquely characterizes $W_{\pm}(\varepsilon)$. Then

$$\begin{aligned}(I + W_{\pm}(\varepsilon))P_{\pm}(\varepsilon) &= I + (\tilde{\pi}_{\pm}(\varepsilon)-\pi_{\pm}) + A^{\pm}(\varepsilon)\\ &= I + \tilde{A}^{\pm}(\varepsilon)\end{aligned} \tag{4.16}$$

where $\tilde{A}^{\pm}(\varepsilon)$ is an analytic function of ε, with values in $OP\mathcal{S}_1^{-1}$, more precisely,

$$\tilde{A}^{\pm}(\varepsilon) = \sum_{j=1}^{\infty} \varepsilon^j \tilde{A}_j^{\pm}, \qquad \tilde{A}_j^{\pm} \in OP\mathcal{H}^{-j}, \tag{4.17}$$

in view of (4.8) and lemma 4.3. Consequently, if we define an operator $\mathfrak{W}$ by

$$\sigma_{\mathfrak{W}}(\pm\lambda)(X,D) = \psi(\varepsilon)W_{\pm}(\varepsilon), \qquad \varepsilon = \lambda^{-\frac{1}{2}}, \tag{4.18}$$

where $\psi(\varepsilon) = 1$ for ε very small, being smooth with small support near $\varepsilon = 0$, we have

$$(I + \mathfrak{W})E = I + \mathfrak{A} \tag{4.19}$$

with

$$\sigma_{\mathfrak{A}}(\pm\lambda)(X,D) = \tilde{A}^{\pm}(\varepsilon), \qquad \varepsilon = \lambda^{\frac{1}{2}} \text{ small}, \tag{4.20}$$

and hence, in light of (4.17),

$$\mathfrak{A} \in OP\Psi^{-1}. \tag{4.22}$$

In order to construct our parametrix, we need only the following information on $\mathfrak{W}$.

Lemma 4.4. We have

$$\mathfrak{W} \in OP\Psi^{\kappa,\infty} \tag{4.23}$$

for some κ.

Proof. To prove (4.23), it suffices to show that, for some κ,

$$\varepsilon^{\kappa} W_{\pm}(\varepsilon) \text{ is analytic in } \varepsilon\text{, with values in } \mathcal{S}_1^{-\infty}. \tag{4.24}$$

We transform the analysis of $W_{\pm}(\varepsilon)$ to a finite dimensional problem. Consider the operators

$$U_{\pm}(\varepsilon) = \pi_{\pm}\,\tilde{\pi}_{\pm}(\varepsilon) + (I-\pi_{\pm})(I-\tilde{\pi}_{\pm}(\varepsilon)). \tag{4.25}$$

It is easy to see that

$$U_{\pm}(\varepsilon): \mathcal{R}(\tilde{\pi}_{\pm}(\varepsilon)) \to \mathcal{R}(\pi_{\pm}) \tag{4.26}$$

where $\mathcal{R}(T)$ denotes the range of the operator T. Also,

$$\begin{aligned} \ker U_{\pm}(\varepsilon) &= \ker\tilde{\pi}_{\pm}(\varepsilon) \cap \mathcal{R}(\pi_{\pm}) \oplus \mathcal{R}(\tilde{\pi}_{\pm}(\varepsilon)) \cap \ker \pi_{\pm} \\ &= \ker U_{\pm}(\varepsilon)^{*}. \end{aligned} \tag{4.27}$$

Furthermore,

$$\begin{aligned} U_{\pm}(\varepsilon) &= I + (I+2\pi_{\pm})\rho_{\pm}(\varepsilon) \\ &= I + K_{\pm}(\varepsilon) \end{aligned} \tag{4.28}$$

where $\rho_{\pm}(\varepsilon) = \tilde{\pi}_{\pm}(\varepsilon) - \pi_{\pm}$, and it is easy to verify from (4.27) that $\ker U_{\pm}(\varepsilon) = \ker U_{\pm}(\varepsilon)^{*}$ is spanned by the +1 and -1 eigenspaces of $K_{\pm}(\varepsilon)$. However, $\|K_{\pm}(\varepsilon)\|$ is small if ε, is, so consequently $U_{\pm}(\varepsilon)$ is invertible for small ε, and hence the map (4.26) is bijective. It follows that

$$W_{\pm}(\varepsilon) = U_{\pm}(\varepsilon)^{-1} X_{\pm}(\varepsilon) U_{\pm}(\varepsilon) \tag{4.29}$$

where $X_{\pm}(\varepsilon)$ is the inverse, on the range of $\pi_{\pm}$, of

$$Y_{\pm}(\varepsilon) = U_{\pm}(\varepsilon) P_{\pm}(\varepsilon) U_{\pm}(\varepsilon)^{-1}. \tag{4.30}$$

Note that $U_\pm(\varepsilon)$ and $U_\pm(\varepsilon)^{-1}$ are analytic functions of ε with values in $OP\mathcal{S}_1^0$.

Now, on the range of $\pi_\pm$, $Y_\pm(\varepsilon)$ is an analytic family of $L_\pm \times L_\pm$ matrices, where $L_\pm$ is the dimension of the range of $\pi_\pm$. Since $P_\pm(\varepsilon)$ is a polynomial in ε, and since $U_\pm(\varepsilon)$ is invertible, uniformly as $\varepsilon \to 0$, it follows that these matrices have determinants which are analytic functions of ε, and, being not identically zero, they vanish to finite order, say order κ, at $\varepsilon = 0$. Thus, Cramer's rule applied to the construction of their inverses yields meromorphic matrix valued functions with poles of order at most κ at $\varepsilon = 0$. This proves the assertion (4.24) and establishes the lemma.

If we denote by $I + \mathcal{B}$ a parametrix for $I + \mathcal{B}$, $\mathcal{B} \in OP\Psi^{-1}$, then we see that

$$(4.31) \qquad (I+\mathcal{B})(I+\mathbb{D})E = I \quad \text{modulo a smoothing operator,}$$

and $(I+\mathcal{B})(I+\mathbb{D}) \in OP\Psi^{\kappa}$. This proves the hypoellipticity of E.

If we recall our original simplifying arguments, we obtain the following more precise version of proposition 4.2.

<u>Proposition 4.5</u>. If $K \in OP\Psi^m$ satisfies the conditions of proposition 4.2, then K is hypoelliptic, with a parametrix of the form

$$(4.32) \qquad L_1 + L_2$$

where

$$(4.33) \qquad L_1 \in OP\Psi^{-m}, \; L_2 \in OP\Psi^{-m+\kappa,\infty}$$

The condition (4.1) is necessary for hypoellipticity, granted the other hypotheses on K of proposition 4.2. Indeed, passing from K to $E \in OP\Psi^0$ as before, we can still obtain the $L_\pm \times L_\pm$ matrix valued analytic function $X_\pm(\varepsilon)$, given by (4.29), making a slight change in the argument in Lemma 4.4 if E is not assumed to be self adjoint. (The self adjointness was used in a minimal way, in (4.27), and can be avoided.) Now, if the hypothesis (4.1) does not hold, then either $X_+(\varepsilon)$ or $X_-(\varepsilon)$ must have determinant identically zero (for small ε). Now, we have the following result.

Lemma 4.6. If $X(\varepsilon)$ is an analytic $L \times L$ matrix valued function, near $\varepsilon = 0$, with determinant identically zero, then there is an analytic function $u(\varepsilon)$ with values in $\mathbb{C}^L$ such that $u(0) \neq 0$ and $X(\varepsilon)u(\varepsilon) \equiv 0$.

Proof. Replacing $X(\varepsilon)$ by $X(\varepsilon)^* X(\varepsilon)$, we can suppose that $X(\varepsilon)$ is self adjoint. Then the result follows from Theorem 1.10, p. 71, of Kato [K1].

So if $X_+(\varepsilon)$ (say) has determinant identically zero, for $|\varepsilon|$ small, define $\tilde{V}_+(\varepsilon)$ to be the orthogonal projection onto such $u(\varepsilon)$, thought of as an analytic function of ε with values in $\mathcal{R}(\pi_+) \subset \mathcal{S}(\mathbb{R}^n) \subset L^2(\mathbb{R}^n)$, and let $V_+(\varepsilon) = U_+(\varepsilon)^{-1}\tilde{V}_+(\varepsilon)U_+(\varepsilon)$, an analytic function of ε with values in $\mathcal{S}_1^{-\infty}$. Then define $S_o \in OP\Psi^{0,\infty}$.

$$\sigma_{S_o}(\lambda)(X,D) = \psi(\lambda)V_+(|\lambda|^{-\frac{1}{2}}) \tag{4.34}$$

where $\psi(\lambda) = 1$ for $\lambda \geq 2M$, $\psi(\lambda) = 0$ for $\lambda \leq M$, M picked sufficiently large. Then we see that S_o is not a smoothing operator, but clearly

$$K\, S_o = 0 \quad \text{mod } OPS^{-\infty}. \tag{4.35}$$

This establishes the following converse to proposition 4.2.

Proposition 4.7. If $K \in OP\Psi^m$ is of the form (4.2) with $\sigma_{K_0}(\pm 1)(X,D)$ elliptic, then the hypothesis (4.1) is necessary as well as sufficient for hypoellipticity of K.

Chapter III: Pseudodifferential operators on contact manifolds

A contact structure on a manifold M, of dimension 2n+1, is given by a line bundle Λ in T^*M which is symplectic in $T^*M \setminus 0$, i.e., the symplectic form on $T^*M \setminus 0$ is nondegenerate acting on tangent vectors to Λ. An alternate characterization is the following: with α a local section of Λ, we demand $\alpha \wedge d\alpha \wedge \cdots \wedge d\alpha \neq 0$, where there are n factors of $d\alpha$. Darboux' theorem implies that any two contact manifolds of the same dimension are locally diffeomorphic via a map preserving the contact structure, i.e., preserving α up to a scalar factor.

The Heisenberg group $\mathbb{H}^n$ forms a convenient standard local model for a contact manifold of dimension 2n+1, as was emphasized by Dynin [D2], following Folland and Stein [F4], who used $\mathbb{H}^n$ as an "infinitesimal" model for constructing kernels on certain CR-manifolds, such as the boundary of a strictly pseudoconvex domain in $\mathbb{C}^{n+1}$. The contact structure we put on $\mathbb{H}^n$ is the line bundle, invariant by right translations, whose fiber over the identity in $\mathbb{H}^n$ is spanned by dt, in the coordinates on $\mathbb{H}^n$ used in Chapter II. This is also the characteristic set of the "Heisenberg Laplacian" $\mathcal{L}_0$, discussed in that chapter.

In the definition of a contact manifold M, associated with the line bundle Λ there is the orthogonal bundle $B \subset TM$, of fiber dimension 2n. A CR-manifold (of maximal complex dimension) with non-degenerate Levi form, is a contact manifold with a family of complex structures on the fibers of B, which satisfy a certain integrability condition. For example, if M is a hypersurface in $\mathbb{C}^{n+1}$, B_x consists of each vector tangent to M at x whose multiple by $i = \sqrt{-1}$ is also in T_xM, which is naturally identified as an $\mathbb{R}$-linear subspace of $\mathbb{C}^{n+1}$. The nondegeneracy condition $\alpha \wedge d\alpha \wedge \cdots \wedge d\alpha \neq 0$ is equivalent to the nondegeneracy of the Levi form.

On such a CR-manifold is a sequence of Kohn Laplacians $\square_b$, some of which are hypoelliptic with loss of one derivative, in the strictly pseudo-convex case. The analysis of such operators via analysis on $\mathbb{H}^n$ provides a nice tool for constructing parametrices of $\square_b$ in such cases. Such parametrices lie in classes of operators determined by the contact structure on M. As concerns the ease of constructing such parametrices, we are fortunate that they lie in classes of operators rather insensitive to the finer CR-structure, since two CR-manifolds need not be locally isomorphic, and classifying these structures may be hopeless. (On the infinitesimal level, one does have the Moser normal form.)

1. Operator classes on contact manifolds.

Since we are interested in a local analysis of pseudodifferential operators on a contact manifold M, we can implement Darboux' theorem, and suppose $M = \mathbb{H}^n$, with the contact structure given by the right invariant line bundle $\Lambda = \text{char } \mathcal{L}_0$, discussed in Chapter II. Thus our operator classes will be obtained from the classes of convolution operators developed in Chapter II, via the machinery fashioned in Chapter I. Recall that we developed in Chapter II the classes $OP\Psi^{m,k}$, $OP\Sigma^m$, $OP\Omega^{m,k}$. Thus, the theory of Chapter I gives us the following operator classes:

$$(1.1) \qquad OP\tilde{\Psi}^m,\ OP\tilde{\Psi}^{m,k},\ OP\tilde{\Sigma}^m,\ OP\tilde{\Omega}^{m,k}.$$

We also make a further study of the classes $OP\tilde{H}^{m,\mu}_{\alpha,\delta}$, developed in Chapter I, when $G = \mathbb{H}^n$. Of these, the class $OP\tilde{\Sigma}^m$ has a transparent behavior. Indeed, the proof of Proposition 1.4 of Chapter I gives immediately:

Proposition 1.1. Locally, and modulo smoothing operators, we have

$$(1.2) \qquad OP\tilde{\Sigma}^m = OPS^m.$$

The other classes require further study. Of course, in light of (2.4) of Chapter II and the discussion following it, we have

$$(1.3)\qquad OP\tilde{\Psi}^m \subset OPS^m_{\frac{1}{2},\frac{1}{2}} \quad \text{if } m \geq 0, \qquad OPS^{m/2}_{\frac{1}{2},\frac{1}{2}} \quad \text{if } m \leq 0,$$

as a corollary to Proposition 1.1 of Chapter I. Similarly, (2.18) of Chapter II implies

$$(1.4)\qquad OP\tilde{\Omega}^{m,k} \subset OPS^{m+\kappa/2}_{\frac{1}{2},\frac{1}{2}}, \qquad \kappa = \max\,(k,0).$$

Now, the whole point of our analysis is not to make much use of the rather weak results (1.3) and (1.4), but to develop symbolic operator calculi for $OP\tilde{\Psi}^m$ and $OP\tilde{\Omega}^{m,k}$. In order to do this, we need to verify the hypotheses that figured in propositions 1.2, 1.3, and 1.5 of Chapter I; these hypotheses are (1.25), (1.26), (1.27), (1.42), (1.43), and (1.46) of Chapter I. We recall them here:

((1.25)) $\mathfrak{X}^m \subset S^m_{\rho\#}$ for some $\rho \in (0,1]$, $m \geq 0$,

((1.26)) $\mathfrak{X}^m \subset S^{m\sigma}_{\rho\#}$ if $m < 0$, for some $\sigma \in (0,1]$,

((1.27)) $A \in OP\mathfrak{X}^m$, $B \in OP\mathfrak{X}^\mu \Rightarrow AB \in OP\mathfrak{X}^{m+\mu}$,

((1.42)) If $p(\xi) \in \mathfrak{X}^m$, then $D^\alpha_\xi p(\xi) \in \mathfrak{X}^{m-\tau|\alpha|}$, for some $\tau \in (0,1]$,

((1.43)) If $K_j \in \mathfrak{X}^{m-\tau j}$, then there exists $K \in \mathfrak{X}^m$ such that $K \sim K_0+K_1+\cdots$,

((1.46)) $p(\xi) \in \mathfrak{X}^m \Rightarrow \overline{p(\xi)} \in \mathfrak{X}^m$.

In the cases of Ψ^m, $\Psi^{m,k}$, $\Omega^{m,k}$, all these results are either immediate from the definition or taken care of explicitly in section 2 Chapter II, except for ((1.42)), which we take care of now.

<u>Proposition 1.2</u>. If $\hat{k}(\tau,y,\eta) \in \Psi^m_o$, then

$$(1.5)\qquad D^j_\tau D^{\gamma_1}_y D^{\gamma_2}_\eta \hat{k}(\tau,y,\eta) \in \Psi^{m-2j-|\gamma_1|-|\gamma_2|}_o.$$

<u>Proof</u>. This is an immediate consequence of differentiating the identity

$$\hat{k}(\tau\lambda,\tau^{\frac{1}{2}}y,\tau^{\frac{1}{2}}\eta) = \tau^{m/2}\,\hat{k}(\lambda,y,\eta)$$

which defines membership in Ψ_o^m.

Proposition 1.3. If $\hat{k} \in \Omega^{m,k}$, then

(1.6) $$D_\tau^j D_y^{\gamma_1} D_\eta^{\gamma_2} \hat{k}(\tau,y,\eta) \in \Omega^{m-j-\frac{1}{2}|\gamma_1|-\frac{1}{2}|\gamma_2|,k-|\gamma_1|-|\gamma_2|}.$$

Proof. This follows immediately from the characterization (2.116) of $\Omega^{m,k}$.

If $K(w)$ is a smooth function, for $w = (t,q,p) \in \mathbb{H}^n$, with values in one of these classes, we denote the symbol of

(1.7) $$\mathbb{K}u(w) = K(w)u(w)$$

by

(1.8) $$\sigma_{\mathbb{K}}(w,\pm\lambda)(X,D) = \pi_{\pm\lambda}(K(w)).$$

Using proposition 1.2 and the machinery of Chapters I and II, we deduce the following:

Proposition 1.4. If $A \in OP\tilde{\Psi}^m$, $B \in OP\tilde{\Psi}^\mu$, then $AB \in OP\tilde{\Psi}^{m+\mu}$. If $C \in OP\tilde{\Psi}^{m+\mu}$ is defined by

(1.9) $$\sigma_C(w,\pm\lambda)(X,D) = \sigma_A(w,\pm\lambda)(X,D)\ \sigma_B(w,\pm\lambda)(X,D)$$

then

(1.10) $$AB - C \in OP\tilde{\Psi}^{m+\mu-1}.$$

To apply proposition 1.3, note that

(1.11) $$\Omega^{\mu-\nu/2,k-\nu} \subset \Omega^{\mu,k}$$

and that a formal series

(1.12) $$\sum_{\nu\geq 0} P_\nu,\quad P_\nu \in OP\Omega^{\mu-\nu/2,k-\nu}$$

asymptotically sums to an element of $OP\Omega^{\mu,k}$. Also

(1.13) $$\Omega^{\mu-\frac{1}{2},k} \subset \Omega^{\mu,k-1}.$$

Hence we have

Proposition 1.5. If $A \in OP\Omega^{m,k}$, $B \in OP\Omega^{\mu,\ell}$, then

(1.14) $$AB \in OP\tilde{\Omega}^{m+\mu,k+\ell}.$$

Furthermore, if C is defined by (1.9), then

(1.15) $$AB - C \in OP\tilde{\Omega}^{m+\mu-\frac{1}{2},k+\ell-1}.$$

We consider now the invariance of $OP\tilde{\Psi}^m$ and $OP\tilde{\Omega}^{m,k}$ under coordinate changes. Let $\varphi\colon \mathbb{H}^n \to \mathbb{H}^n$ be a C^∞ diffeomorphism which preserves the contact structure. If A is an operator on $\mathbb{H}^n$ given by

(1.16) $$Au(w) = \int a(w,wz^{-1})u(z)\,dz, \quad w,z \in \mathbb{H}^n,$$

then $B = \psi^* A\varphi^*$ $(\psi=\varphi^{-1})$ is given by the formula (1.60) of Chapter I:

(1.17) $$Bu(w) = \int b(w,\Psi(w,z,wz^{-1})u(z)\tilde{H}(z)dz,$$

where $b(w,z) = a(\psi(w),z)$ and $\Psi(w,z,y)$ is linear in y, in exponential coordinates. In the formal expansion

(1.18) $$b(w,\Psi(w,z,wz^{-1})) \sim \sum_{\substack{|\gamma|\geq 0 \\ |\sigma|\geq|\gamma|}} C_{\sigma\gamma}(wz^{-1})^{\gamma+\sigma} b_{(\gamma)}(w,\Psi(w)(wz^{-1})),$$

if we set $\gamma = (\gamma_0,\gamma_1,\gamma_2)$ and

(1.19) $$D^\gamma \hat{b}(\tau,y,\eta) = D_\tau^{\gamma_0} D_y^{\gamma_1} D_\eta^{\gamma_2}\,\hat{b}(\tau,y,\eta),$$

then the Fourier transform of the general term in (1.18) is

(1.20) $$C_{\sigma\gamma} D_\tau^{\gamma_0+\sigma_0} D_y^{\gamma_1+\sigma_1} D_\eta^{\gamma_2+\sigma_2}\Big[\tau^{\gamma_0} y^{\gamma_1} \eta^{\gamma_2}\,\hat{b}(w,\Psi(w)^t(\tau,y,\eta))\Big].$$

Now the hypothesis that φ preserves the contact structure on $M = \mathbb{H}^n$ is equivalent to the hypothesis that $\Psi(w)^t = D\psi(w)^t$ preserves the space $\{y=\eta=0\}$. We claim that, in this case, if $A \in OP\tilde{\Psi}_0^m$, then (1.20) is a smooth function of $w \in \mathbb{H}^n$ with values in $\Psi^{m-2\sigma_0-|\alpha_1|-|\sigma_2|} \subset \Psi^{m-|\gamma|}$. This follows from proposition 1.2 if we know that $\hat{b}(w,\Psi(w)^t(\tau,y,\eta)) \in \Psi^m$ for each w, which follows easily from the fact that $\Psi(w)^t$ preserves $\{y=\eta=0\}$. Similarly, using proposition 1.3, one sees that if $A \in OP\tilde{\Omega}^{m,k}$, then (1.20) is a smooth function of w with values in $\Omega^{m-\sigma_0-\frac{1}{2}|\sigma_1|-\frac{1}{2}|\sigma_2|,k-|\sigma_1|-|\sigma_2|} \subset \Omega^{m-\frac{1}{2}|\gamma|,k}$. If the sum in (1.18) is restricted to $0 \leq |\gamma| \leq N$, $|\gamma| \leq |\sigma| \leq 2N$, it is not hard to see that the remainder term represents an arbitrarily smooth kernel if N is sufficiently large. We have the following result.

<u>Proposition 1.6</u>. Let $\varphi\colon \mathbb{H}^n \to \mathbb{H}^n$ be a diffeomorphism which preserves the contact structure, with inverse ψ. If $A \in OP\tilde{\Psi}^m$, then $\psi^* A\varphi^* \in OP\tilde{\Psi}^m$, and if $A \in OP\tilde{\Omega}^{m,k}$, then $\psi^* A\varphi^* \in OP\tilde{\Omega}^{m,k}$.

We now turn our attention to $OP\tilde{H}^{m,\mu}_{\alpha,\delta}$, which was studied for general step 2 nilpotent Lie groups at the end of section 1, Chapter I. In the case $G = \mathbb{H}^n$, $\alpha(s)(t,q,p) = (e^{2s}t,e^{s}q,e^{s}p)$, we can say a few more things. Recall that to say $\mathcal{P}u(w) = P(w)u(w)$ defines $\mathcal{P} \in OP\tilde{H}^{m,\mu}_{\alpha,\delta}$ is equivalent to saying $P(w)u(w) = p_y * u(w)\big|_{y=w}$ and $\hat{p}(w,\zeta) = \hat{p}_w(\zeta)$ has an asymptotic expansion

$$(1.21)\qquad \hat{p}(w,\zeta) \sim \sum_{j\geq 0} a_j(w,\zeta)b_j(w,\zeta)$$

where

$$(1.22)\qquad a_j(w,\zeta) \in \Psi^{m_j}, \quad b_j(w,\zeta) \in \Sigma^{\mu_j} \quad \text{(smooth in } w)$$

with

$$(1.23)\qquad m_j \leq m,\ \mu_j \leq \mu,\ m_j+\mu_j \to -\infty.$$

As shown in Chapter I, proposition 1.17, $\mathcal{P} \in OP\tilde{H}^{m,\mu}_{\alpha,\delta}$ if and only if

$$(1.24)\qquad \mathcal{P} \sim \sum_{j\geq 0} K_j \mathcal{L}_j$$

with

(1.25) $\qquad \mathcal{K}_j \in OP\tilde{\Psi}^{m_j},\ \mathcal{L}_j \in OPS^{\mu_j},\quad m_j \le m,\ \mu_j \le \mu,\ m_j+\mu_j \to -\infty.$

In the case $G = \mathbb{H}^n$, recall that, outside any conic neighborhood of $\Lambda \subset T^*\mathbb{H}^n \setminus 0$, elements of $OP\tilde{\Psi}^m$ belong to OPS^m Thus, microlocally away from Λ, $OP\tilde{\mathcal{H}}^{m,\mu}_{\alpha,\delta}$ coincides with $OPS^{m+\mu}$. In order to understand such operators better near Λ, we can use proposition 2.21 from Chapter II. This yields:

Proposition 1.7. If $K \in OP\tilde{\Psi}^m$, $L \in OPS^\mu$, the symbol of L being supported near Λ, then

$$\sigma_{KL}(w,\pm\lambda)(X,D) \sim \lambda^{\frac{1}{2}m+\mu} \sum_{j\ge 0} \lambda^{-j/2}\, e_j^{\pm}(w,\lambda,X,D) \tag{1.26}$$

where

$$e_0^{\pm}(w,\lambda,x,\xi) = \sigma_K(w,\pm 1)(x,\xi)\ \sigma_L(w,\pm 1)(\lambda^{-\frac{1}{2}}x,\lambda^{-\frac{1}{2}}\xi) \tag{1.27}$$

and generally

$$e_j(w,\lambda,x,\xi) = \sum_{k=1}^{K(j)} p_{kj}^{\pm}(w,\lambda^{-\frac{1}{2}}x,\lambda^{-\frac{1}{2}}\xi)\ a_{kj}^{\pm}(w,x,\xi) \tag{1.28}$$

with

$$p_{kj}^{\pm}(w,x,\xi) \in C^\infty, \text{ compactly supported in } (x,\xi),\quad a_{kj}^{\pm}(w,x,\xi) \in \mathcal{H}_b^{m-j}. \tag{1.29}$$

In a similar fashion we have

$$\sigma_{LK}(w,\pm\lambda)(X,D) \sim \lambda^{\frac{1}{2}m+\mu} \sum_{j\ge 0} \lambda^{-j/2}\, f_j^{\pm}(w,\lambda,X,D) \tag{1.30}$$

where

$$f_0^{\pm} = e_0^{\pm} \tag{1.31}$$

and $f_j^{\pm}(w,\lambda,x,\xi)$ has an expression similar in form to (1.28), (1.29).

In light of the characterization (1.24)-(1.25) of $OP\tilde{\mathcal{H}}^{m,\mu}_{\alpha,\delta}$ we have:

Corollary 1.8. Let the symbol of $\mathcal{P}$ be supported near Λ. Then $\mathcal{P} \in OP\tilde{\mathcal{H}}^{m,\mu}_{\alpha,\delta}$ if and only if $\sigma_{\mathcal{P}}(w,\pm\lambda)(X,D)$ has the form

$$(1.32)\qquad \sigma_{\mu}(w,\pm\lambda)(x,\xi) \sim \lambda^{\frac{1}{2}m+\mu} \sum_{j\geq 0} \lambda^{-j/2}\, e_j(w,\lambda,x,\xi)$$

where $e_j^{\pm}(w,\lambda,x,\xi)$ has the form (1.28), (1.29).

This result makes precise the inclusion

$$(1.33)\qquad OP\tilde{H}^{m,\mu}_{\alpha,\delta} \subset OP\tilde{\Omega}^{\frac{1}{2}m+\mu,m} .$$

We also remark that, in light of the characterization (1.24)-(1.25), proposition 1.6 extends to $A \in OP\tilde{H}^{m,\mu}_{\alpha,\delta}$.

We make some remarks on the formula for a product:

$$(1.34)\qquad AB\, u(x) \sim \sum_{\gamma\geq 0} A^{[\gamma]}(x)\, B_{[\gamma]}(x)\, u(x).$$

We will obtain explicit information on the major terms after the principal one; in particular, with $\gamma = (\gamma_0,\gamma_1,\gamma_2)$, we will look at the cases

$$(1.35)\qquad \gamma = (0,\gamma_1,\gamma_2),\quad |\gamma_1|+|\gamma_2| = 1 \text{ or } 2; \text{ and } \gamma = (1,0,0).$$

We want to understand $A^{[\gamma]}$ on the symbol level. Recall from (1.37) of Chapter I that, if $Au(x) = a*u(x)$, then $A^{[\gamma]}u = a^{[\gamma]}*u$ with

$$(1.36)\qquad \widehat{a^{[\gamma]}}(\xi) = D^{\gamma}\hat{a}(\xi)$$

In case $G = \mathbb{H}^n$, with coordinates (t,q,p) and dual coordinates (τ,y,η), this reads

$$(1.37)\qquad \widehat{a^{[\gamma_0,\gamma_1,\gamma_2]}}(\tau,y,\eta) = D_{\tau}^{\gamma_0} D_y^{\gamma_1} D_{\eta}^{\gamma_2}\, \hat{a}(\tau,y,\eta).$$

Now recall that

$$(1.38)\qquad \hat{a}(\pm\tau,y,\eta) = \sigma_A(\pm\tau)(\pm\tau^{-\frac{1}{2}}y,\tau^{-\frac{1}{2}}\eta)$$

and

$$(1.39)\qquad \sigma_A(\pm\lambda)(x,\xi) = \hat{a}(\pm\lambda,\pm\lambda^{\frac{1}{2}}x,\lambda^{\frac{1}{2}}\xi).$$

It follows easily that

$$(1.40)\qquad \sigma_{A^{[0,\gamma_1,\gamma_2]}}(\pm\lambda)(x,\xi) = \lambda^{-|\gamma|/2}(\pm D_x)^{\gamma_1} D_\xi^{\gamma_2}\,\sigma_A(\pm\lambda)(x,\xi).$$

In particular,

$$(1.41)\qquad D_x^{\gamma_1} = D_{x_j} \Rightarrow \sigma_{A^{[0,\gamma_1,0]}}(\pm\lambda)(X,D) = (\pm 1)\lambda^{-\frac{1}{2}}(D_{x_j}\sigma_A)(\pm\lambda)(X,D)$$

$$= (\pm 1)\lambda^{-\frac{1}{2}}[D_j,\sigma_A(\pm\lambda)(X,D)],$$

and

$$(1.42)\qquad D_\xi^{\gamma_2} = D_{\xi_j} \Rightarrow \sigma_{A^{[0,0,\gamma_2]}}(\pm\lambda)(X,D) = \lambda^{-\frac{1}{2}}(D_{\xi_j}\sigma_A)(\pm\lambda)(X,D)$$

$$= \lambda^{-\frac{1}{2}}[X_j,\sigma_A(\pm\lambda)(X,D)].$$

Equivalently, for right invariant A on $\mathbb{H}^n$, we have, in case (1.41),

$$(1.43)\qquad A^{[0,\gamma_1,0]} = T^{-1}[M_j,A],$$

and, in case (1.42),

$$(1.44)\qquad A^{[0,0,\gamma_2]} = T^{-1}[L_j,A].$$

Recall L_j and M_j are the right invariant vector fields generating the Lie algebra of $\mathbb{H}^n$.

Next, we look at $A^{[1,0,0]}$. From (1.38) we have

$$(1.45)\qquad \sigma_{A^{[1,0,0]}}(\pm\lambda)(x,\xi) = [-\tfrac{1}{2}\lambda^{-1}(x\cdot D_x + \xi\cdot D_\xi) + D_\lambda]\sigma_A(\pm\lambda)(x,\xi).$$

In particular, if $A \in OP\Psi_o^m$,

$$(1.46)\qquad \sigma_{A^{[1,0,0]}}(\pm\lambda)(x,\xi) = -\tfrac{1}{2}\lambda^{-1}(x\cdot D_x+\xi\cdot D_\xi-m)\sigma_A(\pm\lambda)(x,\xi)$$

$$= -\tfrac{1}{2}\lambda^{-1}(r\partial/\partial r-m)\,\sigma_A(\pm\lambda)(x,\xi),$$

where r denotes the radial variable; $r^2 = |x|^2+|\xi|^2$. Note that the principal term in $\sigma_A(\pm\lambda)(x,\xi)$, homogeneous of degree m in (x,ξ), is annihilated by $r\partial/\partial r-m$.

We turn now to $B_{[\gamma]}(y)$, defined by

$$(1.47) \qquad B_{[\gamma]}(y) = (1/\gamma!)\, D_w^\gamma\, B(\exp w)y)\big|_{w=0}, \quad y \in \mathbb{H}^n,$$

in light of the formula (1.34) of Chapter I. If $|\gamma| = 1$, the right side of (1.47) clearly defines a first order differential operator which is a right invariant vector field. Thus,

$$(1.48) \qquad D_x^{\gamma_1} = D_{x_j} \Rightarrow B_{[0,\gamma_1,0]}(y) = (L_j B)(y),$$

and

$$(1.49) \qquad D_\xi^{\gamma_2} = D_{\xi_j} \Rightarrow B_{[0,0,\gamma_2]}(y) = (M_j B)(y).$$

To simplify notation, let $X_\gamma = L_j$ in case (1.48) and $X_\gamma = M_j$ in case (1.49), so we get

$$(1.50) \qquad \gamma_0 = 0,\ |\gamma| = 1 \Rightarrow B_{[\gamma]}(y) = (X_\gamma B)(y).$$

To complete the study of $B_{[\gamma]}(y)$ for $|\gamma| = 1$, note that

$$(1.51) \qquad \gamma = (1,0,0) \Rightarrow B_{[\gamma]}(y) = (TB)(y).$$

Recall that T spans the center of $\mathfrak{h}^n$.

Let us now look at $B_{[0,\gamma_1,\gamma_2]}(y)$ in case $|\gamma_1|+|\gamma_2| = 2$. We can write $(0,\gamma_1,\gamma_2) = \gamma + \gamma'$ in this case where $|\gamma| = |\gamma'| = 1$, so both γ and γ' satisfy the condition (1.50). Note that

$$(1.52) \qquad X_{\gamma'} X_\gamma B(y) = D_{w'}^{\gamma'}\, D_w^\gamma\, B(\exp w)(\exp w')y)\big|_{w,w'=0}.$$

If we use the Campbell-Hausdorff formula

$$(1.53) \qquad (\exp w)(\exp w') = \exp(w+w'+\tfrac{1}{2}[w,w']+\cdots)$$

and plug the right side of (1.52) into

(1.54) $$B(\exp w)y) \sim \sum_{\gamma\geq 0} w^\gamma B_{[\gamma]}(y),$$

we get

(1.55) $$2\, B_{[\gamma+\gamma']}(y) = X_{\gamma'}X_\gamma B(y) - \tfrac{1}{2}[X_{\gamma'},X_\gamma]B(y) = \tfrac{1}{2}(X_{\gamma'}X_\gamma + X_\gamma X_{\gamma'})B(y).$$

Note that, precisely when γ' and γ are complementary, e.g., $D^\gamma_{x,\xi} = D_{x_j}$, $D^{\gamma'}_{x,\xi} = D_{\xi_j}$, we have $[X_{\gamma'},X_\gamma]$ nonzero, and then this commutator is $\pm T$.

2. Subelliptic operators on contact manifolds.

The classes of operators considered in section 1 contain the parametrices of classical pseudodifferential operators on a contact manifold M which are doubly characteristic on the contact line bundle and satisfy a certain condition that will be given below, which will guarantee hypoellipticity with loss of one derivative. As we have seen, we can use a local diffeomorphism preserving the contact structure, and assume M is $\mathbb{H}^n$ with the right invariant contact structure described in the introduction to this chapter.

First consider a second order differential operator on $\mathbb{H}^n$, with nonnegative principal symbol vanishing to precisely second order on $\Lambda \subset T^*\mathbb{H}^n \setminus 0$. Then we have (generalizing (2.70) of Chapter II):

(2.1) $$P = \sum_{j,k} a_{jk}(w)X_jX_k + i\alpha(w)T + B(w,D_w), \qquad w \in \mathbb{H}^n.$$

Here, $a_{jk}(w) = a_{kj}(w)$ is a smooth function of w, forming a positive definite matrix, $\alpha(w) \in C^\infty(\mathbb{H}^n)$, and $B = B(w,D_w)$ is a first order differential operator on $\mathbb{H}^n$ whose principal symbol vanishes on Λ. Hence $B \in OP\tilde{\Psi}^1$. The following result follows immediately from Theorem 2.17 of Chapter II and the machinery of section 1 of this chapter.

Theorem 2.1. Suppose that, for all $y \in \Omega \subset \mathbb{H}^n$, the symbols

(2.2) $$\sigma_{P_2}(\pm 1)(X,D)$$

are elliptic in $OP\mathbb{H}^2$ and invertible on $L^2(\mathbb{R}^n)$, where $P_2(y)$ is the right invariant differential operator

$$(2.3)\qquad P_2(y) = \sum_{j,k} a_{jk}(y)X_jX_k + i\alpha(y)T,$$

which is to say, $\pm\alpha(y)$ avoids for each y the discrete set determined by the spectrum of the second order operator $Q(y,X,D)$ associated with the quadratic form derived from the double sum in (2.3). (See proposition 2.14 of Chapter II.) Then, P, defined by (2.1), is hypoelliptic on Ω, with parametrix in $OP\tilde{\Psi}^{-2}$.

Proof. Let $A \in OP\tilde{\Psi}^{-2}$ be defined by

$$(2.4)\qquad \sigma_A(w,\pm\lambda)(X,D) = \lambda^{-1}\sigma_{P_2(w)}(\pm 1)(X,D)^{-1}.$$

Then proposition 1.5 yields

$$(2.5)\qquad PA = I + R_1, \quad AP = I + R_2, \quad R_j \in OP\tilde{\Psi}^{-1}.$$

From here the standard construction of a parametrix for P goes through.

We remark that the analysis above works if in (2.1) $\alpha(z)$ and B are matrix valued, as long as $a_{jk}(w)$ are scalar. The condition for hypoellipticity is then that all eigenvalues of $\pm\alpha(y)$ avoid the spectrum of $Q(y,X,D)$, for all $y \in \Omega \subset \mathbb{H}^n$. Such a result applies to $\square_b$, the Kohn Laplacian, on the boundary of a strictly pseudoconvex domain, in cases where one has hypoellipciticy with loss of one derivative.

We take the space to explain here the phrase "hypoellipticity with loss of one derivative." If $P \in OPS^m$ is elliptic, we have the regularity result

$$Pu \in H^s_{loc} \Rightarrow u \in H^{s+m}_{loc}.$$

Now, in the case covered by proposition 2.1, since $OP\tilde{\Psi}^{-2} \subset OPS^{-1}_{\frac12,\frac12}$, we have

$$(2.6)\qquad Pu \in H^s_{loc} \Rightarrow u \in H^{s+1}_{loc},$$

rather that $u \in H^{s+2}_{loc}$. This result cannot be improved, as is well known (we will not go into details). This explains the terminology.

We now consider the more general situation, where $P \in OPS^m$ is a classical pseudodifferential operator $M = \mathbb{H}^n$ whose principal symbol is nonnegative and vanishes to exactly second order on $\Lambda \subset T^*\mathbb{H}^n \setminus 0$. We assume the Hessian of p_m transverse to Λ is nondegenerate. For the purpose of constructing a parametrix, we can compose P with an elliptic operator in OPS^{2-m}, and suppose without loss of generality that $P \in OPS^2$. We want to construct a microlocal parametrix belonging to $OP\tilde{\Omega}^{-1,-2}$, under appropriate hypotheses. In the following, Λ^+ denotes a particular connected component of Λ. We will be working microlocally near Λ^+.

First consider the case when P is right invariant on $\mathbb{H}^n$. We can find operators $(A_{jk}) \in OP\Sigma^0$, whose symbol at a point of Λ^+ is a positive definite matrix, and $B \in OP\Sigma^1$, such that

$$(2.7) \qquad P = \sum_{j,k=1}^{2n} A_{jk}X_jX_k + B.$$

Here X_j, $1 \le j \le 2n$, is the set of right invariant vector fields on $\mathbb{H}^n$ given by (2.69) of Chapter II. Now write

$$(2.8) \qquad A_{jk} = a_{jk} + B_{jk}$$

where a_{jk} are real constants and $B_{jk} \in OP\Sigma^0$ have symbol vanishing on Λ^+. Similarly, write

$$(2.9) \qquad B = i\alpha T + \sum_{j=1}^{2n} B_jX_j + B_0, \quad B_j \in OP\Sigma^0, \ 0 \le j \le 2n.$$

Thus we have

$$(2.10) \qquad \begin{aligned} P &= \sum_{j,k} a_{jk}X_jX_k + i\alpha T + \Sigma B_{jk}X_jX_k + \Sigma B_jX_j + B_0 \\ &= P_\alpha + \Sigma B_{jk}X_jX_k + \Sigma B_jX_j + B_0. \end{aligned}$$

Recall that $B_j, B_{jk} \in OP\Sigma^0$, and the symbols of B_{jk} vanish on Λ^+. Since by hypothesis P is elliptic off Λ^+, to construct a parametrix in $OP\Omega^{-1,-2}$, it will suffice to construct a parametrix for (2.10) where B_{jk}, B_j, and B_0 are cut off to have symbols supported in a small conic neighborhood of Λ^+.

The operator

$$P_\alpha = \sum_{j,k} a_{jk} X_j X_k + i\alpha T \tag{2.11}$$

belongs to $OP\Psi_o^2$. Suppose α avoids the appropriate discrete set, so there is

$$P_\alpha^{-1} \in OP\Psi_o^{-2},\ P_\alpha P_\alpha^{-1} = I \text{ microlocally near } \Lambda^+. \tag{2.12}$$

Now look at

$$Q = PP_\alpha^{-1} = I + \Sigma B_{jk} X_j X_k P_\alpha^{-1} + \Sigma B_k X_j P_\alpha^{-1} + B_0 P_\alpha^{-1}. \tag{2.13}$$

Its symbol is

$$\sigma_Q(\pm\lambda)(X,D) = I + \Sigma b_{jk}^{\pm}(\lambda^{-\frac12}X, \lambda^{-\frac12}D) a_{jk}^{\pm}(X,D) + \lambda^{-\frac12}\Sigma b_j^{\pm}(\lambda^{-\frac12}X, \lambda^{-\frac12}D) a_j^{\pm}(X,D) + \lambda^{-1} b_0^{\pm}(\lambda^{-\frac12}X, \lambda^{-\frac12}D) a_0^{\pm}(X,D). \tag{2.14}$$

Here

$$b_{jk}^{\pm}(x,\xi),\ b_j^{\pm}(x,\xi) \in C_o^\infty(\mathbb{R}^{2n}) \tag{2.15}$$

and

$$a_{jk}^{\pm}(x,\xi) \in \mathcal{H}^0,\ a_j^{\pm}(x,\xi) \in \mathcal{H}^{-1}\ \ (j \geq 1),\ a_0^{\pm}(x,\xi) \in \mathcal{H}^{-2}. \tag{2.16}$$

We deduce that

$$\sigma_Q(\pm\lambda)(X,D) = I + r_{\pm}(\lambda,X,D) \tag{2.17}$$

where

$$r_{\pm}(\lambda,x,\xi) \text{ is supported in } |x|+|\xi| \leq C_1\lambda^{\frac12},\ \lambda \geq C, \tag{2.18}$$

and

$$\lambda^k D_\lambda^k r_{\pm}(\lambda,\cdot,\cdot) \text{ is bounded in } \mathcal{S}_1^0, \text{ for } \lambda \geq 1. \tag{2.19}$$

(Since we are working microlocally near Λ^+, we could drop the ± subsrcipts, working only on $\pm\lambda > 0$, but we will instead retain the ± notation.)

Now, in (2.19), we can suppose $r_{\pm}(\lambda,x,\xi)$ is small. In fact, if B_{jk} are cut off sufficiently near Λ, we can suppose, with ε_0 picked small,

$$\sup_{x,\xi} |r_{\pm}(\lambda,x,\xi)| \le C_0(\varepsilon_0+\lambda^{-\frac{1}{2}}), \tag{2.20}$$

and, for $\alpha > 0$,

$$\sup_{x,\xi} |D^{\alpha}_{x,\xi}\, r_{\pm}(\lambda,x,\xi)| \le C_{\alpha}\lambda^{-\frac{1}{2}}. \tag{2.21}$$

The operator norm estimate proved in Appendix A implies

$$\|r_{\pm}(\lambda,X,D)\| \le C(\varepsilon_0+\lambda^{-\frac{1}{2}}). \tag{2.22}$$

Thus, for λ large, $I + r_{\pm}(\lambda,X,D)$ is invertible on $L^2(\mathbb{R}^n)$, so we can define

$$I + s_{\pm}(\lambda,X,D) = (I+r_{\pm}(\lambda,X,D))^{-1}. \tag{2.23}$$

We now undertake to show that $s_{\pm}(\lambda,x,\xi)$ satisfies the conditions (2.19)-(2.21) and a slight modification of (2.18).

Before we analyze $s_{\pm}(\lambda,x,\xi)$, let us define $\sigma_{\pm}(\lambda,x,\xi)$ by the identity

$$1+\sigma_{\pm}(\lambda,x,\xi) = (1+r_{\pm}(\lambda,x,\xi))^{-1} \qquad (\lambda \text{ large}). \tag{2.24}$$

It is easy to see that $\sigma_{\pm}(\lambda,x,\xi)$ satisfies (2.18)-(2.21). In particular, the L^2 operator norm $\|\sigma_{\pm}(\lambda,X,D)\|$ is small, for λ large. Thus we have

$$(I+r_{\pm}(\lambda,X,D))(I+\sigma_{\pm}(\lambda,X,D)) = I - \rho_{\pm}(\lambda,X,D) \tag{2.25}$$

with

$$\|\rho_{\pm}(\lambda,X,D)\| \text{ small, for } \lambda \text{ large.} \tag{2.26}$$

Also $\rho_{\pm}(\lambda,x,\xi) \in \mathcal{S}^{-1}_1$, and more generally

$$\lambda^k D^k_{\lambda}\rho_{\pm}(\lambda,\cdot,\cdot) \text{ is bounded in } \mathcal{S}^{-1}_1, \text{ for } \lambda \text{ large.} \tag{2.27}$$

Furthermore, $\rho_{\pm}(\lambda,x,\xi)$ satisfies the estimate (2.20). Using $\rho_{\pm}(\lambda)$ as shorthand for $\rho_{\pm}(\lambda,X,D)$, we have

$$(I-\rho_{\pm}(\lambda))^{-1} = I + \rho_{\pm}(\lambda) + \cdots + \rho_{\pm}(\lambda)^k + \rho_{\pm}(\lambda)^{\ell}(I-\rho_{\pm}(\lambda))^{-k}\rho_{\pm}(\lambda)^{k-\ell} \tag{2.28}$$

for $0 \le \ell \le k$. It follows that

$$(2.29) \qquad (I-\rho_\pm(\lambda))^{-1} = I + \tau_\pm(\lambda,X,D)$$

with $\lambda^k D_\lambda^k \tau_\pm(\lambda,\cdot,\cdot)$ bounded in $\mathcal{S}_1^{-1}$ for λ large, and $\tau_\pm$ satisfies (2.20), (2.21). Since

$$(2.30) \qquad I + s_\pm(\lambda,X,D) = (I+\sigma_\pm(\lambda,X,D))(I+\tau_\pm(\lambda,X,D))$$

we see that $s_\pm(\lambda,X,D)$ satisfies conditions (2.19)-(2.20).

We proceed to modify $s_\pm(\lambda,x,\xi)$ to obtain (2.18). Note that (2.23) is equivalent to

$$(2.31) \qquad s_\pm(\lambda,X,D) = -r_\pm(\lambda,X,D) - r_\pm(\lambda,X,D)s_\pm(\lambda,X,D).$$

Since $r_\pm(\lambda,X,D)$ satisfies (2.18), we can find $\chi \in C_0^\infty(\mathbb{R}^{2n})$ with $\chi(\lambda^{-\frac{1}{2}}x,\lambda^{-\frac{1}{2}}\xi) = 1$ on supp $r_\pm(\lambda,x,\xi)$, and then

$$(2.32) \qquad r_\pm(\lambda,X,D)\chi(\lambda^{-\frac{1}{2}}X,\lambda^{-\frac{1}{2}}D) - r_\pm(\lambda,X,D) = v_\pm(\lambda,X,D)$$

with $D_\lambda^k v_\pm(\lambda,x,\xi) = O(\lambda^{-\infty})$ in $\mathcal{S}_1^{-\infty}$, for large λ, with a similar result for the products in the reverse order. Thus if we set

$$(2.33) \qquad \tilde{s}_\pm(\lambda,X,D) = s_\pm(\lambda,X,D)\chi(\lambda^{-\frac{1}{2}}X,\lambda^{-\frac{1}{2}}D),$$

formula (2.31) shows that

$$(2.34) \qquad \tilde{s}_\pm(\lambda,X,D) = -r_\pm(\lambda,X,D) - r_\pm(\lambda,X,D)\tilde{s}_\pm(\lambda,X,D) + \tilde{v}_\pm(\lambda,X,D),$$

with $D_\lambda^k\tilde{v}_\pm = O(\lambda^{-\infty})$ in $\mathcal{S}_1^{-\infty}$. This is equivalent to

$$(2.35) \qquad (I+r_\pm(\lambda,X,D))(I+\tilde{s}_\pm(\lambda,X,D)) = I + \delta_\pm(\lambda,X,D)$$

with $\delta_\pm(\lambda,x,\xi) = O(\lambda^{-\infty})$ in $\mathcal{S}_1^{-\infty}$. Thus

$$(2.36) \qquad \sigma_D(\pm\lambda)(X,D) = \delta_\pm(\lambda,X,D)\psi(\lambda)$$

defines a smoothing operator on $\mathbb{H}^n$, where $\psi(\lambda)$ is a smooth cut-off, equal to 1 for λ large, 0 for λ small. Now $\tilde{s}_{\pm}(\lambda,X,D)$ defines an operator $\tilde{S} \in OP\Omega_b^{0,0}$, by

$$(2.37) \qquad \sigma_{\tilde{S}}(\pm\lambda)(X,D) = \tilde{s}_{\pm}(\lambda,X,D),$$

and (2.17) and (2.35) yield

$$(2.38) \qquad PP_{\alpha}^{-1}(I+\tilde{S}) = I+D,$$

with D a smoothing operator on $\mathbb{H}^n$. A similar argument yields a left parametrix for P, and the two are seen to be equal, modulo a smoothing operator, so a two-sided parametrix for P is given by

$$(2.39) \qquad E = P_{\alpha}^{-1}(I+\tilde{S}) \in OP\Omega^{-1,-2}.$$

Given this work on the right invariant case, we can now easily obtain the following general result.

Theorem 2.2. Let $P \in OPS^m$ have nonnegative principal symbol, doubly characteristic on the variety Λ in $T^*\mathbb{H}^n \setminus 0$, satisfying the condition $p_m(z,\zeta) \geq C|\zeta|^{m-2}\,\mathrm{dist}((z,\zeta),\Lambda)^2$. Write P in the form

$$(2.40) \qquad Pu(w) = P(w)u(w)$$

where P(y) is a right invariant (convolution) operator in OPS^m, for each $y \in \mathbb{H}^n$. Construct, for each $y \in \mathbb{H}^n$, the second order differential operator $P_{\alpha}(y)$, according to the prescription (2.7)-(2.11). Suppose that, for each $y \in \Omega \subset \mathbb{H}^n$, the operator $P_{\alpha}(y)$ satisfies the condition for hypoellipciticity (microlocally near Λ^+), given in Theorem 2.1. Then P has a parametrix on Ω (microlocally near Λ^+),

$$(2.41) \qquad E \in OP\tilde{\Omega}^{-m+1,-2},$$

and consequently P is hypoelliptic on Ω (microlocally near Λ^+).

Proof. The argument just given leads to E(y), a smooth function of y with values in $OP\Omega^{-m+1,-2}$, such that $E(y)$ is a parametrix for $P(y)$, for each y. Define $\mathcal{E}_1 \in OP\tilde{\Omega}^{-m+1,-2}$ by

$$(2.42) \qquad \mathcal{E}_1 u(w) = E(w)u(w).$$

Then, by proposition 1.5, especially (1.15), we have

$$(2.43) \qquad \mathcal{E}_1 P = I + R, \qquad R \in OP\tilde{\Omega}^{-\frac{1}{2},-1}.$$

Since $OP\tilde{\Omega}^{-\frac{1}{2},1} \subset OPS^{-\frac{1}{2}}_{\frac{1}{2},\frac{1}{2}}$, R has negative order, so the Neumann series yields a parametrix in $OP\tilde{\Omega}^{0,0}$ for I+R:

$$(2.44) \qquad (I+R)^{-1} = I+S, \qquad S \in OP\tilde{\Omega}^{-\frac{1}{2},-1}.$$

Then a parametrix for P is given by

$$(2.45) \qquad \mathcal{E} = (I+S)\mathcal{E}_1.$$

This finishes the proof.

Since $OP\tilde{\Omega}^{-m+1,-2} \subset OPS^{-m+1}_{\frac{1}{2},\frac{1}{2}}$, we get the regularity results

$$(2.46) \qquad Pu \in H^s_{loc} \Rightarrow u \in H^{s+m-1}_{loc},$$

as in (2.6), i.e., hypoellipticity with loss of one derivative. It is useful to complement this with a couple of other results. If $\mathcal{E} \in OP\tilde{\Omega}^{-m+1,-2}$, then, since $X_j \in OP\Psi^1_o$, we have

$$(2.47) \qquad X_j \mathcal{E} \in OP\tilde{\Omega}^{-m+3/2,-1} \subset OPS^{-m+3/2}_{\frac{1}{2},\frac{1}{2}},$$

and

$$(2.48) \qquad X_j X_k \mathcal{E} \in OP\tilde{\Omega}^{-m+2,0} \subset OPS^{-m+2}_{\frac{1}{2},\frac{1}{2}}.$$

Thus, in addition to (2.46), we have

$$(2.49) \qquad Pu \in H^s_{loc} \Rightarrow X_j u \in H^{s+m-3/2}_{loc} \quad \text{and} \quad X_j X_k u \in H^{s+m-2}_{loc},$$

providing the condition for hypoellipticity of Theorem 2.2 is satisfied.

It is interesting how much simpler it is to handle second order differential operators doubly characteristic on Λ than general pseudodifferential operators. Of course, the Kohn Laplacian $\square_b$ is a differential operator. The operator that arises in the reduction of the $\bar{\partial}$-Neumann problem to the boundary is a pseudodifferential operator $\square^+ \in OPS^1$, characteristic on only one connected component Λ^+ of Λ in $T^*\mathbb{H}^n \setminus 0$. It turns out that there is a $\square^- \in OPS^1$, elliptic on Λ^+, such that

$$\square^-\square^+ = \square_b + R,$$

where $R \in OPS^1$ has principal symbol vanishing on Λ^+; at least this is verified for the $\bar{\partial}$-Neumann problem for (0,1)-forms, in Greiner and Stein [G9]. Thus, microlocally near Λ^+, $R = \sum_{j=1}^{2n} R_j X_j + R_0$ with $R_j \in OPS^0$. This makes it easy to construct a parametrix for $\square^+$, given one for $\square_b$ produced by Theorem 2.1, using less machinery than we developed to prove Theorem 2.2. In fact, in such a case, one has

$$(2.50) \qquad \square_b^{-1}\square^-\square^+ = I + \sum_{j=1}^{2n} \square_b^{-1} R_j X_j + \square_b^{-1} R_0$$

$$= I + S$$

with $S \in OP\tilde{\Omega}^{-\frac{1}{2},1}$. Since S has negative order, the usual method yields a parametrix for $\square^+$ from (2.50).

In this connection, we should mention the following anomaly for $\square_b$ on (0,1)-forms on the boundary of a strictly pseudoconvex domain in $\mathbb{C}^2$ (a three dimensional contact manifold). In this case, the Kohn Laplacian fails to satisfy the condition for hypoellipticity with loss of one derivative. However, microlocally, this failure occurs only on the component Λ^- of Λ, where $\square^+$ is elliptic, and (2.50), microlocalized near Λ^+, still yields a parametrix for $\square^+$ belonging to $OP\tilde{\Omega}^{0,-2}$. Of course, the fact that the $\bar{\partial}$-Neumann problem continues to satisfy the usual subelliptic estimates in this case is well known. We merely point out that construction of a parametrix in this case can still be achieved by a simple application of Theorem 2.1.

It is useful to point out that, in these cases, the parametrix for $\square^+$ belongs to the more refined space:

Proposition 2.3. Under such conditions for hypoellipticity of $\square^+$ as above, if E denotes a parametrix, then

$$E \in OP\tilde{\mathbb{H}}^{-2,1}_{\alpha,\delta} . \tag{2.51}$$

Proof. Since $\square_b^{-1}\,\square^-$ (microlocally near Λ^+) belongs to $OP\tilde{\mathbb{H}}^{-2,1}_{\alpha,\delta}$, and S in (2.50) belongs to $OP\tilde{\mathbb{H}}^{-1,0}_{\alpha,\delta}$, and since

$$E \sim (I-S+S^2-\dots)\square_b^{-1}\,\square^- \quad \text{microlocally near } \Lambda^+,$$

while $E \in OPS^{-1}$ microlocally away from Λ^+, our assertion follows from the algebraic properties of $OP\tilde{\mathbb{H}}^{m,\mu}_{\alpha,\delta}$ proved in Chapter I.

We now give a description of the condition for hypoellipticity of $P \in OPS^m$, with loss of one derivative, given in Theorem 2.2, in terms of the classical symbol of $P \in OPS^m$, doubly characteristic on $\Lambda \subset T^*\mathbb{H}^n \setminus 0$. We shall produce some invariants associated with the symbol of such an operator. These quantities played a role in the papers of Sjöstrand [S6], Boutet de Monvel [B7], Ivrii and Petkov [I1], Hörmander [H9], and others, and they are also discussed in Chapters 13 and 15 of [T2]. For the purposes of this discussion, we shall assume m=2, which involves no loss of generality.

If $P \in OPS^2$ has principal symbol p_2, nonnegative and vanishing to second order on Λ, with nondegenerate transverse Hessian, then, microlocally near any given point of Λ,

$$P_2(x,\xi) = \sum_{j=1}^{\mu} a_j(x,\xi)^2 \tag{2.52}$$

where $a_j(x,\xi)$ are real, homogeneous of degree 1, vanishing on Λ, with da_j

linearly independent at each point of Λ. Hence,

$$P = \sum_{j=1}^{\mu} a_j(x,D)^2 + B(x,D), \qquad B \in OPS^1.$$

On the set Λ, we have

$$(2.53) \qquad B_1(x,\xi) = p_1(x,\xi) - i^{-1} \sum_{|\alpha|=1} a_j^{(\alpha)}(x,\xi) a_{j(\alpha)}(x,\xi)$$

$$= p_1(x,\xi) + \tfrac{1}{2} i \sum_{\nu} (\partial^2 p_2/\partial x_\nu \partial \xi_\nu), \qquad \text{on } \Lambda.$$

The right side of (2.53) is called the subprincipal symbol of P, and is denoted

$$(2.54) \qquad \mathrm{sub}\sigma(P)(x,\xi).$$

It is easy to see that, for $A \in OPS^0$, with principal symbol $a(x,\xi)$,

$$(2.55) \qquad \mathrm{sub}\ \sigma(AP) = \mathrm{sub}\ \sigma(PA) = a(x,\xi)\ \mathrm{sub}\ \sigma(P)(x,\xi) \quad \text{on } \Lambda.$$

Furthermore, if J is an elliptic Fourier integral operator with associated canonical transformation $\mathcal{J}$, then

$$(2.56) \qquad \mathrm{sub}\ \sigma(JPJ^{-1})(\mathcal{J}(x,\xi)) = \mathrm{sub}\ \sigma(P)(x,\xi) \quad \text{on } \Lambda.$$

For two proofs of this, see Chapter 15 of [T2].

Also, at each point $(x_0,\xi_0) \in \Lambda$, we define a Hamilton map F, which is associated to the Hessian of $p_2(x,\xi)$ at (x_0,ξ_0); see the process by which a Hamilton matrix is associated to a quadratic form on a symplectic vector space, as described in proposition 2.14 of Chapter II. It is straightforward that, for the special case

$$(2.57) \qquad P = \sum_{j=1}^{n} \mu_j(L_j^2+M_j^2) + i\alpha T + B = \sum_{j=1}^{n} \mu_j(X_j^2+X_{j+n}^2) + i\alpha T + B,$$

where B is a first order differential operator whose principal symbol vanishes on Λ, we have (with (τ,y,η) denoting dual coordinates to (t,q,p)), on Λ^+,

(2.58) $\quad \text{sub}\ \sigma(P)((t,q,p),(\tau,y,\eta)) = -\alpha|\tau|,$

and

(2.59) $\quad \text{spec}\ i^{-1}F = \{\pm\mu_j|\tau|\} = \{\pm\mu_j^{\#}\}$ (def.)

In particular, if we denote by tr^+F the sum of the positive eigenvalues of $i^{-1}F$ (with multiplicities), we have

(2.60) $\quad tr^+ F = \Sigma\mu_j|\tau|.$

In this case, Theorem 2.2 reduces to a special case of the analysis of the right invariant operators (2.70) of Chapter II. The hypoellipticity condition derived there, in light of (2.58) and (2.59) of this section, is equivalent to the following condition: for all nonnegative integers α_ν, the quantity

(2.61) $\quad \text{sub}\ \sigma(P) + \sum_\nu (2\alpha_\nu+1)\mu_\nu^{\#}$

is nowhere vanishing on Λ; here $\{\mu^{\#}\}$ is the set of positive eigenvalues of $i^{-1}F$. More generally, if P is the second order right invariant differential operator

(2.62) $\quad P = \Sigma\, a_{jk}X_jX_k + i\alpha T + B,$

where B is a first order differential operator whose principal symbol vanishes on Λ, then there is a linear map on $\mathbb{H}^n$, given in fact by a symplectic transformation on $\mathbb{R}^{2n} = \{(q,p)\}$, which takes P to the form (2.57). Such a transformation is a group automorphism of $\mathbb{H}^n$. By the preceding discussion, the quantity (2.61) is invariant under this transformation, so, theorem 2.17 and proposition 2.14 of Chapter II imply that the right invariant differential operator (2.62) is hypoelliptic on $\mathbb{H}^n$ with parametrix in $OP\Psi^{-2}$ if and only if the quantity (2.61) is nonvanishing on Λ.

Now let us pass to general $P \in OPS^2$, $p_2 \geq 0$, vanishing on Λ^+, with nondegenerate transverse Hessian. Then P has the form (generalizing (2.7)):

(2.63) $$Pu(z) = \Sigma A_{jk}(z)X_jX_ku + i\alpha(z)Tu + Bu, \quad z = (t,q,p) \in \mathbb{H}^n.$$

Here, $A_{jk}(z)$ is a smooth function of z with values in $OP\Sigma^0$, $\alpha(z) \in C^\infty(\mathbb{H}^n)$, and $B \in OPS^1$ has principal symbol vanishing on Λ^+. If we work microlocally on Λ^+, which is a ray bundle over $\mathbb{H}^n$, we can write

(2.64) $$A_{jk}(z) = a_{jk}(z) + B_{jk}(z)$$

where $a_{jk}(z) \in C^\infty(\mathbb{H}^n)$ and $B_{jk}(z)$ is a smooth function of z with values in right invariant operators on OPS^1 whose principal symbols vanish on Λ^+. It is easy to show that, for $(z,\zeta) \in \Lambda^+ \subset T^*\mathbb{H}^n \setminus 0$,

$$\text{sub } \sigma(P) = \text{sub } \sigma(\tilde{P})(z,\zeta)$$

where $\tilde{P} \in OPS^2$ is defined by

$$\tilde{P}u(z) = \Sigma a_{jk}(z)X_jX_ku + i\alpha(z)Tu.$$

In fact, we have the stronger statement: at $(z_0,\zeta_0) \in \Lambda^+$,

(2.65) $$\text{sub } \sigma(P)(z_0,\zeta_0) = \text{sub } \sigma(\tilde{P}_{z_0})(z_0,\zeta_0)$$

where $\tilde{P}_{z_0}$ is the right invariant differential operator:

(2.66) $$\tilde{P}_{z_0} u(z) = \Sigma a_{jk}(z_0)X_jX_ku(z) + i\alpha(z_0)Tu(z).$$

Furthermore, the Hamilton matrices for P and $\tilde{p}_{z_0}$ have the same eigenvalues at $(z_0,\zeta_0) \in \Lambda^+$. Consequently, Theorem 2.2 implies the following result.

Theorem 2.3. Let $P \in OPS^2$ have principal symbol $p_2 \geq 0$, vanishing to second order on the variety $\Lambda^+ \subset T^*\mathbb{H}^n \setminus 0$ defining the contact structure for $\mathbb{H}^n$. Let $(z_0,\zeta_0) \in \Lambda^+$ and suppose that, for all nonnegative integers α_ν, the quantity (2.61) is nonvanishing at (z_0,ζ_0). Then P is microlocally hypoelliptic on a conic neighborhood of (z_0,ζ_0), with microlocal parametrix in $O\tilde{P}\Omega^{-1,-2}$.

The microlocal hypoellipticity of P, with loss of one derivative, and parametrix in the operator class of the form (2.121) of Chapter II, is a special case of a more general result proved by Sjöstrand [S6] and Boutet de Monvel [B7], and Boutet de Monvel and Treves [B13]:

Theorem 2.4. Let $P \in OPS^2$ have principal symbol $p_2 \geq 0$ on $T^*\Omega \setminus 0$, vanishing to second order on a conic set Σ, assumed to be a symplectic submanifold of $T^*\Omega \setminus 0$. Assume a nondegenerate transverse Hessian. Then P is hypoelliptic with loss of one derivative provided that, at each point Σ, for all nonnegative integers α_ν, the quantity (2.61) is nonvanishing.

These authors also prove that (2.61) is necessary for hypoellipticity with loss of one derivative, as well as sufficient, granted the other hypotheses of the theorem. In [B7], a parametrix E is produced with

$$E \in OPS^{-2,-2}(\Omega,\Sigma) \subset OPS^{-1}_{\frac{1}{2},\frac{1}{2}}(\Omega).$$

As mentioned in Chapter II, section 2, if $(\Omega,\Sigma) = (\mathbb{H}^n,\Lambda)$, we have

$$OP\tilde{\Omega}^{-1,-2} \subset OPS^{-2,-2}(\mathbb{H}^n,\Lambda).$$

Hypoellipticity of P with loss of one derivative, under more general conditions, when Σ need not be symplectic, is studied in detail in [B10], [H8], and other places, but we will not go into it here.

We do want to show that "half" of Theorem 2.4 follows from Theorem 2.3.

Proposition 2.5. Suppose 2n is the codimension of Σ in $T^*\Omega \setminus 0$. Then, if

$$k = \dim \Omega \leq 2n+1,$$

Theorem 2.3 implies Theorem 2.4.

Proof. The proof will utilize the following result in symplectic geometry. If Σ_j are conic submanifolds of $T^*\Omega_j \setminus 0$, if $\dim \Omega_1 = \dim \Omega_2$ and $\dim \Sigma_1 = \dim \Sigma_2$, and if Σ_j are both symplectic submanifolds, then, given $p_j \in \Sigma_j$, there exists a homogeneous symplectic diffeomorphism from a conic neighborhood U_1 of p_1 to a conic neighborhood U_2 of p_2, taking $\Sigma_1 \cap U_1$ onto $\Sigma_2 \cap U_2$. First consider the case

$$\dim \Omega = 2n+1 = \operatorname{codim} \Sigma + 1. \tag{2.67}$$

In this case, with $\Sigma_1 = \Omega$, $\Sigma_1 = \Sigma$, we can take $\Omega_2 = \mathbb{H}^n$, $\Sigma_2 = \Lambda$, and get Σ locally symplectically equivalent to Λ. If we implement this local canonical transformation by an elliptic Fourier integral operator J, we see that JPJ^{-1} satisfies the hypotheses of Theorem 2.3 on its principal symbol. Since the quantity (2.61) is invariant under conjugation by J, it follows that the condition for hypoellipticity of P in Theorem 2.4 is equivalent to the condition for hypoellipticity of JPJ^{-1} given in Theorem 2.3. This establishes proposition 2.5 in case $\dim \Omega = \operatorname{codim} \Sigma + 1$. Now suppose that

$$\operatorname{codim} \Sigma + 1 = \dim \Omega + \ell, \quad \ell > 0. \tag{2.68}$$

Let us form $\tilde{\Omega} = \Omega \times R^\ell$, and let $\tilde{P} = P \otimes I$. If

$$Pu = f \tag{2.69}$$

we have

$$\tilde{P}\tilde{u} = \tilde{f} \tag{2.70}$$

where $\tilde{u}(x,y) = u(x)$, $x \in \Omega$, $y \in \mathbb{R}^\ell$, and $\tilde{f}$ is similarly defined. Note that $WF(\tilde{u})$ and $WF(\tilde{f})$ are contained in $\{\eta=0\}$, if (ξ,η) are dual to $(x,y) \in \Omega \times \mathbb{R}^\ell$. $\tilde{P}$ is not quite a pseudodifferential operator, but its symbol is smooth near $\{\eta=0\}$, so we can treat it as a pseudodifferential operator, when analyzing (2.70). Note that, if $\tilde{\Sigma}$ is the characteristic variety of P, then, near $\{\eta=0\}$, we have

$$\tilde{\Sigma} = \{(x,y,\xi,\eta):(x,\xi) \in \Sigma\}.$$

Thus $\operatorname{codim} \tilde{\Sigma} = \operatorname{codim} \Sigma = \dim \Omega + \ell - 1 = \dim \tilde{\Omega} - 1$. Note that, if (2.61) is nonvanishing for P at $(x_o,\xi_o) \in \Sigma$, then it is nonvanishing for $\tilde{P}$ at

$(x_0,y_0,\xi_0,0)$. Thus the previous argument applies, to give hypoellipticity for $\tilde{P}$, which implies hypoellipticity for P. This proves proposition 2.6.

Note that, generally, $2 \leq \operatorname{codim} \Sigma \leq 2 \dim \Omega$. If $2 \leq \operatorname{codim} \Sigma \leq \dim \Omega - 2$, Theorem 2.4 can be proved using analysis on the group $\mathbb{H}^n \times \mathbb{R}^\ell$, the cartesian product of a Heisenberg group and an abelian group. We will say a little more about this in Chapter IV.

3. Heat equations and spectral asymptotics for subelliptic operators.

Suppose M is a compact contact manifold, of dimension 2n+1. Let P be a negative self adjoint second order differential operator, whose principal symbol p_2 is ≥ 0, vanishing to exactly second order on $\Lambda \subset T^*M \setminus 0$, the span of the contact form on M. Assume

$$(3.1) \qquad |\text{sub}\, \sigma(P)| < \text{tr}^+F \quad \text{on} \quad \Lambda$$

where F is the Hamilton map of p_2, and tr^+F is the sum of the positive eigenvalues of $(1/i)F$. It follows from Theorem 2.4 that P is hypoelliptic. Note that $P \in OP\tilde{\Psi}^2$; its parametrix belongs to $OP\tilde{\Psi}^{-2}$. In particular, P has compact resolvent, since

$$(3.2) \qquad (\lambda - P)^{-1} : L^2(M) \to H^1(M),$$

so P has discrete spectrum. We aim to study the spectral asymptotics of P, by means of a study of the "heat semigroup" e^{tP} and an asymptotic analysis of $\text{tr}\, e^{tP}$ as $t \to 0$. Menikoff and Sjöstrand [M8], have made such studies in more general contexts. However, the method used here is shorter and simpler. It also has the advantage that we produce a complete asymptotic expansion for $\text{tr}\, e^{tP}$ as $t \downarrow 0$, whereas Menikoff and Sjöstrand produce only the principal term. We note that Metivier [M9] has results on spectral asymptotics for P of the form ΣX_j^2, X_j vector fields, using analysis on Nilpotent Lie groups. Also Fefferman and Phong [F3] have estimates on the spectrum of subelliptic operators in very general contexts. We also mention Iwasaki and Iwasaki [I2].

Locally, on an open set $U \subset M$, mapped diffeomorphically to an open set in $\mathbb{H}^n$, preserving the contact form, we can write

$$(3.3) \qquad Pu(x) = P(x)u(x)$$

where, for each y, $P(y)$ is a right invariant second order differential operator on $\mathbb{H}^n$. For y fixed, we have analyzed the semigroup $e^{-tP(y)}$ on $\mathbb{H}^n$ in Chapter II, section 3. We have (for $y \in U$, $(x,z) \in \mathbb{H}^n$)

$$(3.4) \qquad e^{tP(y)}\delta_o(s,z) = e^{tb(y)}\, k_t(y;s,z), \qquad t > 0,$$

where

$$(3.5) \qquad k_t(y;s,z) = k_t^{Q(y)}(s+it\alpha(y),z),$$

the right side of (3.5) being given by (3.61)-(3.63) of Chapter II. Here we suppose

$$P(y) = \Sigma\, a_{jk}(y)X_jX_k + i\alpha(y)T + b(y).$$

$Q(y)$ is the quadratic form determined by $(a_{jk}(y))$, as in (2.71) of Chapter II. Note that

$$(3.6) \qquad k_t(y;s,z) = t^{-n-1}\, k_1(y;s/t,z\sqrt{t})$$

and $k_1(y;\cdot,\cdot)$ is a smooth function of y with values in $\mathcal{S}(\mathbb{H}^n)$. Thus, for $(s,z) \neq (0,0)$, $k_t(y;s,z)$ vanishes to infinite order as $t \downarrow 0$. Define $k^{\#}(y;t,s,z)$ by

$$(3.7) \qquad k^{\#}(y;t,s,z) = \begin{cases} k_t(y;s,z) & \text{for } t \geq 0, \\ 0 & \text{for } t < 0. \end{cases}$$

Note that (3.6) yields

$$(3.8) \qquad k^{\#}(y;rt,rs,r^{\frac{1}{2}}z) = r^{-n-1}\, k^{\#}(y;t,x,z).$$

Also, $k^{\#}$ is smooth for $(t,s,z) \neq (0,0,0)$. Let us set

(3.9) $k_{\#}(y;t,s,z) = \chi(t,s,z)\, k^{\#}(y;t,s,z)$

where χ is a compactly supported cut-off, equal to 1 near (0,0,0). Then $k_{\#}$ is a smooth function of y with values in $\hat{\mathcal{H}}(G,\alpha,-2)$, where

(3.10) $G = \mathbb{R} \times \mathbb{H}^n,\ \alpha(\rho)(t,s,z) = (r^2t, r^2s, rz),\quad r = e^{\rho},\ t \in \mathbb{R},\ (s,z) \in \mathbb{H}^n.$

Note that $\alpha(\rho)$ expands volumes by a factor of $r^{2n+4} = e^{(2n+4)\rho}$. Let us denote

(3.11) $\mathfrak{X}_0^m = \mathcal{H}(G,\alpha,m)$

with G and α given by (3.10), and let $\mathfrak{X}^m$ consist of asymptotic sums of elements of $\mathfrak{X}_0^{m-j}$, j=0,1,2,.... We see that, if K_0 is defined by

(3.12) $K_0u(t,x) = k_b(t,x;\cdot)*u(t,x), \qquad (t,x) \in \mathbb{R} \times \mathbb{H}^n,$

where $k_b = e^{tb(y)}k_{\#}$. Then

(3.13) $K_0 \in OP\tilde{\mathfrak{X}}^{-2}.$

In view of the results of Chapter I, we have

(3.14)
$$(\partial/\partial t - P)K_0u(t,x) = (I-\mathcal{R})u,$$
$$\mathcal{R} \in OP\tilde{\mathfrak{X}}^{-1}.$$

Note that both K_0 and $\mathcal{R}$ have the <u>evolution property</u>: if v(t,x) is defined on $\mathbb{R} \times \mathbb{H}^n$ and vanishes for t < T, so do K_0v and $\mathcal{R}v$. Also, so does any iterate $\mathcal{R}^k$. Proposition 1.8 of Chapter I implies

(3.15) $\mathcal{R}^k \in OP\tilde{\mathfrak{X}}^{-k},$

and

(3.16) $K_0\mathcal{R}^k \in OP\tilde{\mathfrak{X}}^{-2-k}.$

Thus $(\partial/\partial t - P)$ has a left parametrix

(3.17) $$\mathbb{K} \sim K_0 + K_0R + K_0R^2 + \cdots$$

$$= K_0 + K_1 + K_2 + \cdots,$$

satisfying the evolution property, and similar considerations show $\mathbb{K}$ is a right parametrix. Now, if

(3.18) $$\mathbb{K}v(t,x) = K(t,x;\cdot)*v(t,x), \qquad (t,x) \in \mathbb{R} \times \mathbb{H}^n,$$

the convolution being on $\mathbb{R} \times \mathbb{H}^n$, we have $K(t,x;t',x')$ independent of t and

(3.19) $$e^{tP}\,\delta_p(x) = K(0,x;x^{-1}p), \qquad t > 0,$$

modulo an error smooth in $\overline{\mathbb{R}}^+ \times M$. Thus

(3.20) $$\operatorname{tr} e^{tP} = \int_M K(0,x;t,0)\, d\operatorname{vol}(x) + A(t)$$

where $A \in C^\infty(\overline{\mathbb{R}}^+)$. Now, by (3.17), upon rearrangement, we can write

(3.21) $$K(t,x;t',x') \sim \sum_{j\geq 0} K_j(t,x;t',x')$$

where $K_j(t,x;t',x')$ is smooth in (t,x) and approximately homogeneous of degree $-2n-2+j$ in (t',x'), with respect to the dilations (3.10). According to proposition 1.9 of Chapter I, subtracting a smooth function from K_j, we can suppose

(3.22) $$K_j(t,x;r^2t',r^2s,rz) = r^{-2n-2+j}\, K_j(t,x;t',x,z)$$

provided

(3.23) $$j < 2n+2.$$

In particular, modulo a smooth function of t,

(3.24) $$K_j(0,x;t,0) = t^{-n-1+\frac{1}{2}j}\, K_j(0,x;1,0), \qquad \text{for } j < 2n+2.$$

For $j \geq 2n+2$, say $j = 2n+1+k$, $(k \geq 1)$, we can apply proposition 1.9 of Chapter I to $D^k_{t'}K_j$, the kernel of an element of $OP\tilde{X}^{-2-j+2k}$, and conclude

(3.25) $D^k_{t'} K_j(0,x;rt',0) = r^{-\frac{1}{2}-\frac{1}{2}k} D^k_{t'} K_j(0,x;t',0).$

Integrating, we see that K_j must be given by (a smooth function plus)

(3.26) $K_j(0,x;t,0) = C_j\ t^{(k-1)/2} K_j(0,x;1,0)$ if $k \geq 0$ is even,

and

(3.27) $C'_j\ t^{(k-1)/2} K_j(0,x;1,0) + C''_j(x)\ t^{(k-1)/2} \log t,$ if $k \geq 1$ is odd,

where

(3.28) $k = j - 2n - 1 \geq 0.$

From (3.20), we have our conclusion:

Theorem 3.1. If P is a second order differential operator on a compact contact manifold satisfying the hypotheses of the first paragraph of this section, then

(3.29) $$\operatorname{tr} e^{tP} \sim t^{-n-1}(A_0 + B_0 t^{\frac{1}{2}} + A_1 t + B_1 t^{3/2} + \cdots + A_n t^n + B_n t^{n+\frac{1}{2}} + \cdots)$$
$$+ (A'_{n+1} \log t + A'_{n+2}\ t \log t + \cdots).$$

It is possible that all the coefficients of the log terms are zero. This would be the case if P were an elliptic differential operator; see Duistermaat and Guillemin [D1] for a discussion in that case.

The asymptotic expansion (3.29) together with Karamata's Tauberian Theorem (for a proof, see p. 341 of [T2]) implies the following result on the eigenvalue asymptotics for P. Suppose

(3.30) $\operatorname{spec} P = \{-\lambda_1, -\lambda_2, \ldots\},\ \lambda_j \geq 0,$

and

(3.31) $N(\lambda) = \operatorname{card}\{-\lambda_j \in \operatorname{spce} P : \lambda_j < \lambda\}.$

Corollary 3.2. Under the hypotheses of theorem 3.1, we have

(3.32) $$\lim_{\lambda\to\infty} \lambda^{-n-1} N(\lambda) = C = A_0/(n+1)\Gamma(n+1).$$

Our construction of a parametrix for $\partial/\partial t - P$ on $\mathbb{R} \times M$ could have proceeded along lines more directly analogous to the construction of a parametrix for P in section 2, utilizing a symbol calculus on $\mathbb{R} \times \mathbb{H}^n$. The infinite dimensional irreducible unitary representations of this group are of the form

$$(3.33) \qquad \pi_{\sigma,\pm\lambda}(t,s,q,p) = e^{i\sigma t}\pi_{\pm\lambda}(s,q,p),\ \sigma \in \mathbb{R},\ \lambda \in (0,\infty),$$

where $t \in \mathbb{R}$, $(s,q,p) \in \mathbb{H}^n$. If $k \in \mathcal{E}'(\mathbb{R} \times \mathbb{H}^n)$, $Ku = k*u$,

$$(3.34) \qquad \begin{aligned} \pi_{\sigma,\pm\lambda}(k) &= \hat{k}(\sigma,\pm\lambda,\pm\lambda^{\frac{1}{2}}X,\lambda^{\frac{1}{2}}D) \\ &= \sigma_K(\sigma,\pm\lambda)(X,D), \end{aligned}$$

which implies

$$(3.35) \qquad \hat{k}(\sigma,\pm\tau,y,\eta) = \sigma_K(\sigma,\pm\tau)(\pm\tau^{-\frac{1}{2}}y,\tau^{-\frac{1}{2}}\eta).$$

We are interested in $\hat{k}$, smooth on $(\sigma,\lambda,y,\eta) \neq 0$ and homogeneous:

$$(3.36) \qquad \hat{k}(\tau\sigma,\tau\lambda,\tau^{\frac{1}{2}}y,\tau^{\frac{1}{2}}\eta) = \tau^{m/2}\,\hat{k}(\sigma,\lambda,y,\eta),\ \tau > 0.$$

This homogeneity is equivalent to

$$(3.37) \qquad \sigma_K(\tau\sigma,\pm\tau\lambda)(x,\xi) = \tau^{m/2}\,\sigma_K(\sigma,\pm\lambda)(x,\xi), \quad \tau > 0.$$

It would be convenient to have a result parallel to proposition 2.2 of Chapter II, characterizing when $\sigma_K(\sigma,\pm\lambda)$ given such that (3.37) holds gives $\hat{k}$, via (3.35), which is C^∞ on the complement of the origin. In setting down (3.4),(3.5), to construct the parametrix K of $\partial/\partial t - P$, we have exploited the explicit calculation from section 3 of Chapter II, rather than such a general theory of harmonic analysis and pseudodifferential operators on $\mathbb{R} \times \mathbb{H}^n$, whose construction we will not pursue here, but merely mention to the reader as one sort of direction in which to extend the theory developed in this chapter and the preceding one.

We point out that Melin [M4] has developed an operator calculus and parametrix construction (but not a symbol calculus) in a context which contains $\mathbb{R} \times M$ here as a special case.

Note added: R. Beals has informed the author that joint work nearing completion with P. Greiner and N. Stanton has produced the expansion (3.29). They have been able to show the log terms are absent, and they also note a cancellation effect showing all the B_j in (3.29) are zero. Also joint work of N. Stanton and D. Tartakoff treats the heat equation for $\Box_b$ on a strictly pseudoconvex CR manifold with a "Levi metric" which involves assuming the Hamilton map has the smallest possible number (two) of distinct eigenvalues at each point on the characteristic set.

4. Szegö operators, Toeplitz operators, and hypoellipticity with loss of two derivatives.

Let M be a compact contact manifold, with contact line bundle Λ, and Λ_+ one of the two connected components of $\Lambda \setminus 0$. As in section 3, let P be a self adjoint second order differential operator (or pseudodifferential operator, $P \in OPS^2$), whose principal symbol P_2 is ≥ 0, vanishing on Λ but with Hessian nondegenerate transverse to $\Lambda \setminus 0$.

We have seen that P is hypoelliptic with loss of one derivative provided that, for all nonnegative integers α_ν, the quantity

$$(4.1) \qquad \text{sub}\, \sigma(P) + \sum_\nu (2\alpha_\nu + 1)\mu_\nu$$

is nowhere vanishing on $\Lambda \setminus 0$. See Theorem 2.3. Here we shall consider a case where this condition is violated uniformly on Λ_+ (we will suppose it holds on the other connected component Λ_-, for definiteness). We will suppose (4.1) vanishes with $\alpha_\nu = 0$:

$$(4.2) \qquad \text{sub}\, \sigma(P) + \text{tr}^+ F = 0 \text{ on } \Lambda_+.$$

The best known case when this happens is when P is the Kohn Laplacian $\Box_b^{(0)}$ on functions (0-forms) on the boundary of a strongly pseudoconvex domain $\mathcal{O}$. In this case, if S is the Szegö projector onto the space of boundary values of functions holomorphic on $\mathcal{O}$, $\square_b^{(0)} S = S\, \square_b^{(0)} = 0$. Now there are holomorphic functions in $\mathcal{O}$ with singularities at any given point of $\partial\mathcal{O}$, hence singular elements of the range of S, so $\square_b^{(0)}$ is certainly not hypoelliptic. However, many zero order perturbations of $\square_b^{(0)}$ turn out to be hypoelliptic.

One of our goals in this section is to analyze the structure of the Szegö projector for $\square_b^{(0)}$ mentioned above. Also we will examine the relationship between an operator $P \in OPS^2$ having a Szegö operator and certain zero order perturbations of P being hypoelliptic with loss of two derivatives. First, some definitions. An operator $P \in OPS^2$ will be said to have a Szegö operator S and complementary parametrix E if the following three conditions hold:

$$(4.3) \qquad E \in OPS^{-1}_{\frac12,\frac12}, \quad S \in OPS^{0}_{\frac12,\frac12},$$

$$(4.4) \qquad EP + S \sim I,$$

$$(4.5) \qquad SP \sim 0,$$

where we say

$$(4.6) \qquad A \sim B \Longleftrightarrow A - B \in OPS^{-\infty}.$$

If these conditions hold microlocally on some conic open set $\Gamma \subset T^* M \setminus 0$, we say P has a Szegö operator S and complementary parametrix E microlocally on Γ. One might want to weaken the hypotheses (4.3); actually, when we show such operators exist, we will want to obtain them in more precise operator classes, such as $OP\tilde{\Psi}^{-2}$ and $OP\tilde{\Psi}^{0,\infty}$, respectively, in some cases. It is also useful to weaken hypotheses (4.4) and (4.5) to

$$(4.7) \qquad EP + S = I + R_1, \quad SP = R_2, \quad R_j \in OPS^{-\varepsilon}_{\frac12,\frac12},$$

for some $\varepsilon > 0$. A pair (S,E) satisfying (4.3) and (4.7) will be called a weak Szegö operator with complementary parametrix.

The following result generalizes the argument of Stein [S7], which we discussed in Chapter II; see (2.63)-(2.67) of Chapter II.

Proposition 4.1. Let $P \in OPS^2$ satisfy the hypotheses of the first paragraph (of this section) and (4.2). Suppose P has a weak Szegö operator S with complementary parametrix E. Suppose $P_1 = P + A$, with $A \in OPS^0$. Then P_1 is hypoelliptic with loss of two derivatives, provided the principal symbol of A is nonvanishing on Λ_+.

Proof. Since P is assumed to satisfy away from Λ_+ the condition for hypoellipticity with loss of one derivative, we can suppose S is essentially supported in any given conic neighborhood of Λ_+. Let $B \in OPS^0$ be a parametrix for A, microlocally near Λ_+. Then BS is well defined, mod $OPS^{-\infty}$, and we have

$$(4.8) \qquad (E + BS)P_1 = (E+BS)(P+A) = I - S + R_1 + BR_2 + BSA + EA = I - R_3$$

where

$$(4.9) \qquad R_3 = S - BSA - R_1 - BR_2 - EA \in OPS^{-\varepsilon}_{\frac{1}{2},\frac{1}{2}}$$

It follows that the Neumann expansion yields a parametrix for $I-R_3$:

$$(4.10) \qquad F \sim I + R_3 + R_3^2 + \cdots \in OPS^0_{\frac{1}{2},\frac{1}{2}},$$

so

$$(4.11) \qquad F(E+BS)P_1 = I \bmod OPS^{-\infty}.$$

Thus P_1 has a left parametrix in $OPS^0_{\frac{1}{2},\frac{1}{2}}$, and the proposition is proved.

We will next show that, when a Szegö operator S and complementary parametrix E exist (mocrolocally), they are microlocally unique. One possible use of this is to reduce the global construction of Szegö operators to microlocal constructions.

Proposition 4.2. Let $P \in OPS^2$ be self adjoint. Suppose (4.3)-(4.5) hold on an open conic set $\Gamma \subset T^*M \setminus 0$. Then, on Γ,

$$S \sim S^* \sim S^2, \tag{4.12}$$

and, if $S_1 \in OPS^0_{\frac{1}{2},\frac{1}{2}}$ satisfies

$$S_1 P \sim 0, \tag{4.13}$$

then

$$S_1 \sim S_1 S. \tag{4.14}$$

Furthermore, if there is $E_1 \in OPS^{-1}_{\frac{1}{2},\frac{1}{2}}$ such that

$$E_1 P + S_1 \sim I, \tag{4.15}$$

then

$$S_1 \sim S, \tag{4.16}$$

and, provided E and E_1 are normalized so

$$ES \sim 0 \tag{4.17}$$

and $E_1 S_1 \sim 0$, then

$$E \sim E_1. \tag{4.18}$$

Finally, granted the normalization (4.17), we have

$$E \sim E^*. \tag{4.19}$$

<u>Proof</u>. If $P = P^*$, then (4.4) implies

$$PE^* + S^* \sim I. \tag{4.20}$$

If we multiply on the left by S, we get

$$S \sim SS^*.$$

Since this implies $S^* \sim SS^*$, this gives (4.12). If we multiply (4.20) on the left by S_1, hypothesis (4.13) implies

$$S_1 \sim S_1 S^*,$$

and, in light of (4.12), (4.14) follows. Now, if (4.15) holds, then, as with S, we have $S_1 \sim S_1^* \sim S_1^2$, and $S \sim SS_1$. Combining with (4.14), we have

$$S_1 \sim S_1^* \sim S^* S_1^* \sim SS_1 \sim S_1,$$

so (4.16) is proved.

Note that, if (4.4) holds, it also holds with E replaced by E(I-S), so the normalization (4.17) may be assumed. If E_1 enjoys the same sort of normalization, since $S \sim S_1$, we have

$$(E-E_1)P \sim 0,$$

and hence, by (4.14), in conjunction with (4.17) and its analogue for E_1,

$$E-E_1 \sim (E-E_1)S \sim 0,$$

so (4.18) is proved. Finally,

$$(I-S)E^* \sim EPE^* \sim E(I-S),$$

and hence, if $ES \sim SE^* \sim 0$, we obtain (4.19).

Suppose $P \in OPS^2$ is a self adjoint operator on a compact manifold, M, satisfying the conditions of this section, such that one can construct microlocally on a conic neighborhood of each point of $T^*M \setminus 0$ a Szegö operator S with complementary parametrix E. By the uniqueness of proposition 4.2, it follows that there exist global operators $S \in OPS^0_{\frac12,\frac12}$, $E \in OPS^{-1}_{\frac12,\frac12}$ on M, such that (4.4) and (4.5) hold. It follows that (4.12), (4.17), and (4.19) hold globally on M. One can replace S by $\frac12(S+S^*)$, altering it by a smoothing operator, and achieve $S = S^*$. From the relation $S^2 - S \in OPS^{-\infty}$ we deduce that S can be altered further by an element of $OPS^{-\infty}$, so that

$$S^* = S \text{ and } S = S^2, \tag{4.21}$$

i.e., S is an orthogonal projection. We still have $SP \sim PS \sim 0$, of course. However, it is not necessarily true that S can be taken to be the orthogonal projection onto the kernel of P. If the orthogonal projection onto ker P does satisfy the condition to be a Szegö operator, we say is is the Szegö projector (associated with P). The orthogonal projection (on $L^2(\partial\mathcal{O})$) onto the space of boundary values of functions holomorphic on a strongly pseudoconvex domain $\mathcal{O}$, is indeed a Szegö projector associated with the Kohn Laplacian $\square_b^{(0)}$. We will discuss this in a moment. On the other hand, it follows from an example of Nirenberg [N5] that there is a P on a three dimensional M such that ker P consists of constants, and the orthogonal projection fails to satisfy (4.3)-(4.5).

We turn now to the analysis of the orthogonal projection S on $L^2(\partial\mathcal{O})$ onto the space of boundary values of functions holomorphic in $\mathcal{O}$, given $\mathcal{O} \subset \mathbb{C}^k$ strongly pseudoconvex. We begin with some observations of Boutet de Monvel and Sjöstrand [B12]. We first consider the case $\dim \partial\mathcal{O} \geq 5$, so $k \geq 3$. We have the $\bar{\partial}_b$ complex; see, e.g., Krantz [K9]; and S is the orthogonal projection onto ker $\square_b^{(0)}$, where $\square_b^{(0)} = \bar{\partial}_b^* \bar{\partial}_b$ on 0-forms. More generally, we have $\square_b^{(j)} = \bar{\partial}_b^* \bar{\partial}_b + \bar{\partial}_b \bar{\partial}_b^*$ on sections of certain vector bundles E_j; $\bar{\partial}_b$ maps sections of E_j to sections of E_{j+1}. Note that

$$(4.22) \qquad \bar{\partial}_b \square_b^{(0)} = \bar{\partial}_b \bar{\partial}_b^* \bar{\partial}_b = \square_b^{(1)} \bar{\partial}_b .$$

Now, as long as $k \geq 3$, as is well known, the Kohn Laplacian $\square_b^{(1)}$ on E_1 satisfies the criterion for hypoellipticity with loss of one derivative, as in Theorem 2.1 and Theorem 2.3 of this chapter. Consequently, there exists

$$(4.23) \qquad L_1 \in OP\tilde{\Psi}^{-2}$$

such that

$$(4.24) \qquad L_1 \square_b^{(1)} = \square_b^{(1)} L_1 = I - S_1, \quad S_1 L_1 = L_1 S_1 = 0, \; L_1 = L_1^*,$$

where S_1 is the orthogonal projection onto the kernel of $\square_b^{(1)}$, which is a finite dimensional space of smooth sections of E_1. Thus $\bar\partial_b \bar\partial_b^* L_1$ is the orthogonal projection onto the image of $\bar\partial_b$ and $\bar\partial_b^* L_1 \bar\partial_b$ is the orthogonal projection onto the orthogonal complement of ker $\bar\partial_b$. Hence S is given by

$$(4.25) \qquad S = I - \bar\partial_b^* L_1 \bar\partial_b \in OP\tilde\Psi^0;$$

this formula for S was noted in [B12]. If we set

$$(4.26) \qquad E = \bar\partial_b^* L_1^2 \bar\partial_b (I-S) \in OP\tilde\Psi^{-2},$$

we see that

$$(4.27) \qquad S\,\Box_b^{(0)} = 0,\ E\,\Box_b^{(0)} = I-S,\ ES = 0,$$

so S is a Szegö operator for $\Box_b^{(0)}$ with complementary parametrix E.

The formula (4.25) for S is not wholly satisfactory. For example, it hides the fact that S belongs to $OPS^{-\infty}$ outside any conic neighborhood of Λ_+, which is an automatic consequence of the fact that $S\,\Box_b^{(0)} = 0$ together with the hypoellipticity of $\Box_b^{(0)}$ away from Λ_+. We will proceed to analyze S as an element of the smaller operator class $OP\tilde\Psi^{0,\infty}$, and produce an explicit formula for its leading term. This will produce an analysis of the leading singularity of S alternate to, and in some respects simpler than, that of Boutet de Monvel and Sjöstrand [B13].

In this case, in local coordinates identifying a patch of M with a neighborhood of the identity in $\mathbb{H}^n$ (dim M = 2n+1), we have $\Box_b^{(0)}$ of the form

$$(4.28) \quad Pu = P(x)u(x) = \Sigma a_{jk}(x)X_jX_ku + i\alpha(x)Tu + \Sigma b_j(x)X_ju + c(x)u, \quad x \in \mathbb{H}^n.$$

We have shown that there exists a Szegö operator $S \in OP\tilde\Psi^0$ and complementary parametrix $E \in OP\tilde\Psi^{-2}$. Say

$$(4.29) \qquad Su(x) = S(x)u(x), \quad Eu(x) = E(x)u(x),$$

where S(x) is a smooth function on $\mathbb{H}^n$ with values in Ψ^0, etc. According to Proposition 4.2, the complete symbols of S and E are uniquely determined by the properties $SP \sim 0$, $EP+S \sim I$. Let us proceed to derive their principal symbols.

Now the hypothesis (4.2), which is satisfied by $P = \square_b^{(0)}$, implies that, for each $x_o \in \mathbb{H}^n$, the right invariant operator

$$(4.30) \qquad P_2(x_o) = \Sigma a_{jk}(x_o)X_jX_k + i\alpha(x_o)T$$

fails to satisfy the condition for hypoellipticity that $\sigma_{P_2(x_o)}(+1)(X,D)$ be invertible. Indeed, this self adjoint operator has a one dimensional kernel, varying smoothly with x_o, and consisting of elements of $\mathcal{S}(\mathbb{R}^n)$. Let $\pi_o(x_o)$ denote the L^2-orthogonal projection onto this kernel. Now define

$$(4.31) \qquad S_0 \in OP\tilde{\Psi}^{0,\infty}$$

by

$$(4.32) \qquad \sigma_{S_0}(x,\lambda)(X,D) = \begin{cases} \pi_o(x) & \text{for } \lambda > 0 \\ 0 & \text{for } \lambda < 0 \end{cases} .$$

Similarly define

$$(4.33) \qquad E_0 \in OP\tilde{\Psi}^{-2}$$

by the requirement that $\sigma_{E_0}(x,\lambda)(X,D)$ be the inverse of $\sigma_{P_2}(x,\lambda)(X,D)$ for $\lambda < 0$, and for $\lambda > 0$ it should annihilate the range of $\pi_o(x)$ and should invert σ_{P_2} on the kernel of $\pi_o(x)$. Now since

$$(4.34) \qquad \begin{aligned} P(x) &= P_2(x) + \Sigma b_j(x)X_j + c(x) \\ &= P_2(x) + B(x), \end{aligned}$$

with $B(x)$ smooth in Ψ^1, we see from the symbol calculus of section 1 that S_0 and E_0 must be the principal parts of S and E:

$$(4.35) \qquad S-S_0 \in OP\tilde{\Psi}^{-1}, \quad E-E_0 \in OP\tilde{\Psi}^{-3}.$$

That S_0 belongs to $OP\tilde{\Psi}^{0,\infty}$ and not merely to $OP\tilde{\Psi}^0$ is a hint of the following:

(4.36) $\quad S \in OP\tilde{\Psi}^{0,\infty}.$

To see this, note that the relation $S\,\square_b^{(0)} = 0$ together with the fact that $\square_b^{(0)}$ is hypoelliptic off Λ_+ implies S belongs to $OPS^{-\infty}$ microlocally outside any conic neighborhood of Λ_+. Thus (4.36) is a consequence of the following.

Lemma 4.3. Let $T \in OP\tilde{\Psi}^m$, and suppose that, away from any conic neighborhood of $\Lambda_+ \subset T^*M \setminus 0$, T belongs to $OPS^{-\infty}$. Then $T \in OP\tilde{\Psi}^{m,\infty}$.

Proof. If $Tu(z) = T(z)u(z)$, the hypothesis implies each right invariant $T(z_o) \in OP\Psi^m$ belongs to $OPS^{-\infty}$ outside Λ_+. Set

$$T(z) \sim \sum_{j \geq 0} T_j(z)$$

with

$$\sigma_{T_j}(z,\pm\lambda)(X,D) = \lambda^{(m-j)/2}\, \tau_j^{\pm}(z,X,D).$$

We have

$$\sigma_{T_j}(z,-\lambda)(X,D) = 0 \quad \text{for } \lambda > 0,$$

for all j. Now, the characterization of $OP\Psi_o^{m-j}$ given by Proposition 2.2 of Chapter II implies $\sigma_{T_j}(z,+\lambda)(X,D) \in OP\mathcal{S}_1^{-\infty}$, $\lambda > 0$, for each j, and this completes the proof.

A general element of $OP\tilde{\Psi}^{m,\infty}$ belongs to $OPS^{-\infty}$ away from $\Lambda \setminus 0 = \Lambda_+ \setminus 0 \cup \Lambda_- \setminus 0$. If $T \in OP\tilde{\Psi}^{m,\infty}$ actually belongs to $OPS^{-\infty}$ away from Λ_+, we will say

(4.37) $\quad T \in OP\tilde{\Psi}_+^{m,\infty}.$

Let us return to specifying the complete asymptotic expansion of S and E for $\Box_b^{(0)}$. We have $S_0 \in OP\tilde{\Psi}_{o+}^{0,\infty}$ and $E_0 \in OP\tilde{\Psi}_o^{-2}$ such that

(4.39) $$S_oP = R_0 \in OP\tilde{\Psi}_+^{1,\infty}, \quad E_0P + S_0 - I = Q_0 \in OP\tilde{\Psi}^{-1}.$$

We will produce by induction

(4.40) $$S_j \in OP\tilde{\Psi}_{o+}^{-j,\infty}, \quad E_j \in OP\tilde{\Psi}_o^{-2-j},$$

such that

(4.41) $$(S_0+\cdots+S_k)P = R_k \in OP\tilde{\Psi}_+^{1-k,\infty}$$

and

(4.42) $$(E_0+E_1+\cdots+E_k)P + (S_0+\cdots+S_k) - I = Q_k \in OP\tilde{\Psi}^{-1-k}.$$

We also want $ES \sim 0$, which is equivalent to

(4.43) $$(E_0+\cdots+E_k)(S_0+\cdots+S_k) = W_k \in OP\tilde{\Psi}^{-3-k}.$$

Suppose we have $S_0,\ldots,S_k$ and $E_0,\ldots,E_k$. We will specify S_{k+1} and E_{k+1}. Let $R_k^{\#}$ and $Q_k^{\#}$ denote the principal parts of R_k and Q_k, belonging, respectively, to $OP\tilde{\Psi}_o^{1-k,\infty}$ and $OP\tilde{\Psi}_o^{-1-k}$. Similarly let $W_k^{\#}$ denote the principal part of W_k. We have

$$R_k^{\#}\, u(x) = R_k^{\#}(x)u(x), \quad x \in \mathbb{H}^n,$$

where $R_k^{\#}(y)$ is smooth in y with values in $OP\Psi_o^{1-k,\infty}$, and similar formulas for $Q_k^{\#}$, etc. Now for (4.41) to hold with k replaced by k+1, it is necessary and sufficient that, for each $y \in \mathbb{H}^n$,

(4.44) $$S_{k+1}(y)P_2(y) = - R_k^{\#}(y).$$

We note that a necessary and sufficient condition for there to exist $S_{k+1} \in OP\tilde{\Psi}_o^{-k-1,\infty}$ such that this holds is that the quantity

(4.45) $$\pi_1(R_k^{\#}(y)S_0(y)) = \sigma_{R_k^{\#}(y)S_0(y)}(+1)(X,D)$$

vanish for each y. For a general P of the form (4.28) this might conceivably be a nontrivial condition. However, for $P = \square_b^{(0)}$, we know that S and E exist, in the operator classes $\widetilde{OP\Psi}_+^{0,\infty}$ and $\widetilde{OP\Psi}^{-2}$, so the vanishing of (4.45) is automatic. It follows that

$$(4.46) \qquad S_{k+1}(y) = -R_k^{\#}(y)E_0(y) + \alpha_{k+1}(y)\, S_0(y),$$

where $\alpha_{k+1}(y)$ is a smooth function of y with values in $OP\Psi_o^{-k-1}$ which remains to be determined. In view of the formulas of Chapter I, section 1, applied to the symbol of the product S_0P_2, we have

$$(4.47) \qquad R_0^{\#}(y) = \sum_{|\gamma_1|+|\gamma_2|=1} S_0^{[0,\gamma_1,\gamma_2]}(y)\; P_{2[0,\gamma_1,\gamma_2]}(y) + \Sigma b_j(y)S_0(y)X_j.$$

Specifying $\alpha_{k+1}(y)$ will be a byproduct of the study of E_{k+1}, to which we turn. For (4.42) to hold with k replaced by k+1, it is necessary and sufficient that, for each y,

$$(4.48) \qquad E_{k+1}(y)P_2(y) = -S_{k+1}(y) - Q_k^{\#}(y).$$

Now, for (4.48) to be solvable for $E_{k+1}(y)$, it is necessary and sufficient that

$$(4.49) \qquad \pi_1([S_{k+1}(y) + Q_k^{\#}(y)]S_0(y))$$

vanish. This vanishing is equivalent to

$$\pi_1(\alpha_{k+1}(y))\pi_o(y) = -\pi_1(Q_k^{\#}(y))\pi_o(y),$$

so we can replace (4.46) by

$$(4.50) \qquad S_{k+1}(y) = -R_k^{\#}(y)E_0(y) - Q_k^{\#}(y)S_0(y).$$

With this granted, we have

$$(4.51) \qquad \begin{aligned} E_{k+1}(y) &= -[S_{k+1}(y)+Q_k^{\#}(y)]E_0(y) + \beta_{k+1}(y)S_0(y) \\ &= [R_k^{\#}(y)E_0(y)-Q_k^{\#}(y)]E_0(y) + \beta_{k+1}(y)S_0(y). \end{aligned}$$

Here, the factor $\beta_{k+1}(y)$ is a smooth function of y with values in $OP\Psi_{o}^{-k-3}$ that remains to be specified. This specification is a result of (4.43), with k replaced by k+1, which requires

$$E_0(y)S_{k+1}(y) + E_{k+1}(y)S_0(y) + W_k^{\#}(y) = 0.$$

If we plug in (4.50) and (4.51), we see this is equivalent to

$$\beta_{k+1}(y)S_0(y) = - W_k^{\#}(y) + E_0(y)R_k^{\#}(y)E_0(y) + E_0(y)Q_k^{\#}(y)S_0(y). \tag{4.52}$$

Again this apparently nontrivial identity is a consequence of the existence of S and E in $OP\tilde{\Psi}^0$ and $OP\tilde{\Psi}^{-2}$, and it uniquely specifies $\beta_{k+1}(y)S_0(y)$, since the other terms in (4.52) have already been specified. We can rewrite (4.51) as

$$E_{k+1}(y) = [R_k^{\#}(y)E_0(y)-Q_k^{\#}(y)]E_0(y) + E_0(y)[R_k^{\#}(y)E_0(y)+Q_k^{\#}(y)S_0(y)] - W_k^{\#}(y). \tag{4.53}$$

To summarize, for $k \geq 3$, we have proved the following result. We note that a result along these lines was suggested by Dynin [D2].

<u>Theorem 4.4</u>. Let $\mathcal{O} \subset \mathbb{C}^k$ be a strongly pseudoconvex domain, with smooth boundary $M = \partial\mathcal{O}$. Then the Szegö projector S, the orthogonal projection of $L^2(M)$ onto the space of boundary values of holomorphic functions, is an operator belonging to $OP\tilde{\Psi}_{+}^{0,\infty}$, whose complete symbol is given by (4.32) and (4.50).

The construction of the Szegö projector for $\Box_b^{(0)}$ based on (4.25) does not quite work for k=2, i.e., dim $\partial\mathcal{O} = 3$. In that case, $\Box_b^{(1)}$ fails to be hypoelliptic, so we cannot arrange (4.24). However, the hypoellipticity of $\Box_b^{(1)}$ fails only on Λ_-, not on Λ_+. Thus we can pick $L_1 \in OP\tilde{\Psi}^{-2}$ to be a parametrix for $\Box_b^{(1)}$ microlocally away from Λ_-, and if we define S_a and E_a by (4.25) and (4.26), we see that

$$S_a \in OP\tilde{\Psi}^0, \; E_a \in OP\tilde{\Psi}^{-2}$$

and

$$S_a \square_b^{(0)} = A, \quad E_a \square_b^{(0)} + S_a = I - B, \; E_a S_a = C,$$

where $A \in OP\tilde{\Psi}^{2,\infty}$, $B \in OP\tilde{\Psi}^0$, and $C \in OP\tilde{\Psi}^{-2,\infty}$ belong to $OPS^{-\infty}$ outside any conic neighborhood of Λ_-. In analogy with (4.37), and with Lemma 4.3 in mind, we write

$$A \in OP\tilde{\Psi}_-^{2,\infty}, \quad B \in OP\tilde{\Psi}_-^{0,\infty}, \quad C \in OP\tilde{\Psi}_-^{-2,\infty}.$$

The important point is that, microlocally on a conic neighborhood of Λ_+, S_a and E_a furnish a Szegö operator and complementary parametrix for $\square_b^{(0)}$. Since $\square_b^{(0)}$ has a parametrix in $OP\tilde{\Psi}^{-2}$ away from Λ_+, this shows that in the case dim $\partial\mathcal{O} = 3$, we have a Szegö operator S' and complementary parametrix for $\square_b^{(0)}$ on $\partial\mathcal{O}$, and the complete symbol of the Szegö operator is still specified by (4.32) and (4.50). Once one has this result, it is possible to deduce that the Szegö projector in this case is equal to S' and $OPS^{-\infty}$, using Kohn's estimates for the $\bar{\partial}$-Neumann problem. This argument is carried out in Boutet de Monvel and Sjöstrand [B12]. It relies heavily on $M = \mathcal{O}$ being embedded in $\mathbb{C}^k$ as the boundary of a strongly pseudoconvex domain. The Szegö projector for more general strongly pseudoconvex CR-manifolds has been studied by Kohn [K6]. We also mention another approach to the analysis of the Szegö projector, of Kerzman and Stein [K3]. The Szegö projector is related to the Bergman kernel function; see [B12], and also the paper [B1] of Beals, Fefferman, and Grossman for a description of this and also of Fefferman's original analysis of the Bergman kernel function.

Let us now take a closer look at what sort of operator S was produced by the preceding construction. First look at S_0, defined by (4.32). Take the case where

(4.54) $\quad P_2(y) = \mathcal{L}_0 + inT.$

Then $\pi_o = \pi_o(y)$ is the orthogonal projection onto the lowest eigenspace of $H = -\Delta+|x|^2$, which is spanned by $\pi^{-n/2}\, e^{-|x|^2/2}$. From the formula (1.13) of Chapter II relating an operator to its Weyl symbol, we readily obtain

$$\pi_o = \pi_o(X,D), \quad \pi_o(x,\xi) = c_n\, e^{-|x|^2-|\xi|^2} . \tag{4.55}$$

Now if we write

$$S_0\, u(x) = s_o * u(x) \tag{4.56}$$

in this case, the formulas

$$\hat{s}_o(\tau,y,\eta) = \pi_o(\tau^{-\frac{1}{2}}y,\tau^{-\frac{1}{2}}\eta) = c_n\, e^{-(|y|^2+|\eta|^2)/\tau} \qquad (\tau > 0) \tag{4.57}$$

and

$$\hat{s}_o(-\tau,y,\eta) = 0 \qquad (\tau > 0)$$

give

$$s_o(t,q,p) = c_n'(\partial/\partial t)^n\, [\tfrac{1}{4}|q|^2+\tfrac{1}{4}|p|^2+it]^{-1}, \tag{4.58}$$

a formula derived in Folland and Stein [F4].

In the more general case

$$P_2(y) = \Sigma a_{jk}(y)X_jX_k + i\alpha(y)T, \tag{4.59}$$

under the hypothesis on α which follows from (4.2), we see that $\pi_o = \pi_o(y)$ is the orthogonal projection onto the lowest eigenspace of $Q(X,D) = Q(y,X,D)$, where

$$Q(x,\xi) = \sum_{j,k=1}^{2n} a_{jk}(y)\chi_j\chi_k \tag{4.60}$$

and

$$\chi_j = x_j,\ \chi_{j+n} = \xi_j,\ 1 \le j \le n. \tag{4.61}$$

If we choose a symplectic basis of $\mathbb{R}^{2n}$ to diagonalize $Q(x,\xi)$ and exploit metaplectic covariance, we see that $\pi_0 = \pi_0(X,D)$ with

(4.62) $$\pi_0(x,\xi) = C(Q)\ e^{-\psi(Q,x,\xi)},$$

where $\psi(Q,x,\xi)$ is a positive quadratic form in (x,ξ), given explicitly by

(4.63) $$\psi(Q,x,\xi) = Q(A_Q^{-1}\zeta,\zeta),\ \zeta = (x,\xi).$$

Compare formula (A.31) of Appendix A, or formulas (3.51) and (3.60) of Chapter II. Now, if we write $S_0(y)$ as a convolution operator:

(4.64) $$S_0(y)u(x) = s_{oy} * u(x).$$

we have

(4.65) $$\hat{s}_{oy}(\tau,y,\eta) = \begin{cases} C_n(Q)\ e^{-\psi(Q,y,\eta)/\tau} & \text{if } \tau > 0, \\ 0 & \text{if } \tau < 0, \end{cases}$$

and hence

(4.66) $$s_{oy}(t,q,p) = \int_0^\infty \iint e^{-\psi(Q,y,\eta)/\tau}\ e^{iy\cdot q + i\eta\cdot p + i\tau t}\ dy\ d\eta\ d\tau.$$

Now the Fourier transform of a Gaussian is another Gaussian:

(4.67) $$\iint e^{-\psi(Q,y,\eta)/\tau}\ e^{iy\cdot q+i\eta\cdot p}\ dy\ d\eta = C'(Q)\tau^n\ e^{-\psi^{\#}(Q,q,p)\tau},$$

where, if we use the natural coordinates on $\mathbb{R}^{2n}$ to represent the quadratic form $\psi(Q,x,\xi)$ as a $2n \times 2n$ matrix,

$$\psi(Q,x,\xi) = \zeta\cdot\Psi_Q\zeta,$$

we have

(4.68) $$\psi^{\#}(Q,q,p) = \tfrac{1}{4}\ z\cdot\Psi_Q^{-1}\ z,\ z = (q,p).$$

Hence the distribution s_{oy} from (4.64) is given by

(4.69) $$s_{oy}(t,q,p) = C''(Q)\ (\partial/\partial t)^n[\psi^{\#}(Q,q,p) + it]^{-1}.$$

Using (4.65) or (4.69), it is possible to analyze S_0 as a Fourier integral operator with complex phase, associated with a certain positive almost complex canonical relation $\mathfrak{C}$, and of order zero:

$$(4.70) \qquad S_0 \in I^0(M,M;\mathfrak{C}).$$

See Melin and Sjöstrand [M5] for the theory of Fourier integral operators with complex phase. Inductively from (4.50) one can obtain

$$(4.71) \qquad S_j \in I^{-j/2}(M,M;\mathfrak{C}).$$

Thus S is a Fourier integral operator "adapted" to the contact structure of $M = \partial\mathcal{O}$, in the sense of [B11], Appendix A. We omit the details of passing from (4.65) to (4.70) and (4.71), which do not differ essentially from the model case analysis given in [B11] and [B12].

It may be tempting to ask if the order of the elements in the expansion of S actually drops by integers rather than half integers. To see that this does not necessarily happen, consider

$$(4.72) \qquad P = \mathcal{L}_0 + inT + iX_1 - 1.$$

We will construct the Szegö operator for this right invariant operator on $\mathbb{H}^n$. With $S \sim S_0 + S_1 + \cdots$, $S_j \in OP\tilde{\Psi}^{-j,\infty}_{+o}$, we see that S_0 must be the operator defined by (4.55)-(4.58), the Szegö operator for $\mathcal{L}_0+inT = P_2$. If S_1 exists, it must satisfy the identity (4.44), which in this case becomes

$$(4.73) \qquad S_1P_2 = -iS_0X_1 .$$

The condition for solvability,

$$(4.74) \qquad \pi_1(S_0X_1S_0) = 0,$$

is seen to hold; elements of the (one dimensional) range of $\pi_1(S_0)$ are even; applying X_1 makes them odd, hence orthogonal to the range of $\pi_1(S_0)$. In this case, we have

$$(4.75) \qquad S_1 = -iS_0X_1E_0 - iE_0X_1S_0.$$

If S_2 exists, it must satisfy

(4.76) $$S_2P_2 = -iS_1X_1 - S_0$$
$$= -S_0X_1E_0X_1 - E_0X_1S_0X_1 - S_0.$$

Note that $\pi_1(\mathcal{L}_0+inT)$ acts as the identity operator on the range of $\pi_1(X_1S_0)$, so $S_0X_1E_0X_1S_0 = S_0X_1^2S_0$. Now

$$\pi_1(S_0X_1^2S_0) = -\gamma\pi_1(S_0)$$

with

$$\gamma = \left(\int_{\mathbb{R}^n} x_1^2\, e^{-|x|^2}dx\right)/\left(\int_{\mathbb{R}^n} e^{-|x|^2}\, dx\right) = 1.$$

This guarantees that when the right side of (4.76) is multiplied on the right by S_0, the product vanishes. Indeed, the exact Szegö operator for (4.72) is given by

(4.77) $$S = S_0+S_1+S_2,\quad S_j \in OP\tilde{\Psi}_+^{-j,\infty}\quad (S_j \in I^{-j/2}(M,M;\mathbb{C})),$$

and the term S_1, given by (4.75), is not zero.

It might be objected that the operator (4.72) is not related to a $\bar{\partial}_b$ complex. However, P happens to be unitarily equivalent to $\mathcal{L}_0$+inT; a simple calculation shows

(4.78) $$\mathcal{L}_0+inT+iX_1-1 = e^{\frac{1}{2}iT^{-1}X_{n+1}}(\mathcal{L}_0+inT)e^{-\frac{1}{2}iT^{-1}X_{n+1}}.$$

Microlocally near Λ, $T^{-1}X_{n+1}$ is a pseudodifferential operator in OPS^0, and so is $\exp(\pm\frac{1}{2}iT^{-1}X_{n+1})$. Thus the operator (4.72) is microlocally conjugate to an operator of the form $\square_b^{(0)}$ near the characteristic set. In particular, the Szegö projector (4.77) is microlocally conjugate to the Szegö projector S_0, given by (4.56)-(4.58).

The construction (4.31)-(4.57) of the complete symbol of the Szegö operator S and complementary parametrix E for $P = \square_b^{(0)}$ used the foreknowledge that S and E existed (being given by (4.25) and (4.26)), in the classes $OP\tilde{\Psi}^0$ and $OP\tilde{\Psi}^{-2}$, respectively. One could consider more general situations where this construction could be carried out, and we plan to pursue this in a future publication.

Given such a Szegö operator S acting on $\mathcal{D}'(M)$, one defines a Toeplitz operator (or order m) as an operator of the form

$$(4.79) \qquad T = SQS, \qquad Q \in OPS^m(M).$$

As shown by Boutet de Monvel and Sjöstrand [B12], any two Szegö operators are microlocally conjugate, via a Fourier integral operator, so microlocally, one is reduced to studying (4.79) where $S = S_0$ is the Szegö projector for $\mathcal{L}_0+inT$ on $\mathbb{H}^n$ (or any other local model). We will not develop the basic theory of Toeplitz operators as a consequence of the operator calculus for $OP\tilde{\Psi}^{2m,\infty}$. Instead we refer to Boutet de Monvel and Guillemin [B11], where the basic theory is worked out from a different point of view. We mention the following three results, proved in the first chapters of [B11].

Theorem A. If $Q \in OPS^m$ has principal symbol vanishing on Λ_+, then SQS is a Toeplitz operator of order m-1.

Theorem B. Any Toeplitz operator (4.79) can be written in the form $T = SQ'S$ where $Q' \in OPS^m$ commutes with S. If T is self adjoint, Q' can be taken self adjoint.

Theorem C. If T_1 and T_2 are Toeplitz operators, of order m_1 and m_2, respectively, then T_1T_2 is a Toeplitz operator of order m_1+m_2.

We remark that Theorem C follows from Theorem B, and that Theorem A is used to prove Theorem B. Theorem A might seem surprising, in light of (4.77), and we will say a little more about it. This phenomenon is strongly related to a result of R. Howe, which we will discuss in Appendix B, which sets up a natural isomorphism between the space OPH_b^{2m} of pseudodifferential operators on $\mathbb{R}^n$ and the space of Toeplitz operators of order m on the unit sphere in $\mathbb{C}^n$. Theorem B sets one up to study spectral asymptotics for $T_Q = SQS$ in case Q is elliptic and self adjoint, by studying $e^{isT_Q} = S\, e^{isQ'}S$. Most of [B11] is devoted to studying the composition of Fourier integral operators like $e^{isQ'}$ and operators like S, and applying this to obtain deep information on the behavior of the spectrum of T_Q.

Let us take a look at the symbol of a Toeplitz operator (4.79), as an element of $OP\tilde{\Psi}^{2m,\infty}$, in the case $S = S_0$, the Szegö projector for $\mathcal{L}_0+inT$ on $\mathbb{H}^n$. Set

$$\sigma_{S_0}(\lambda)(X,D) = \pi_o(X,D), \ \lambda > 0. \tag{4.80}$$

As we have seen, $\pi_o(x,\xi)$ is given by (4.55). If $Q \in OP\tilde{\Sigma}_o^m$, then, from our operator calculus, we have, for $w \in \mathbb{H}^n$,

$$\sigma_{QS_0}(w,\lambda)(X,D) \sim \lambda^m \sum_{k\geq 0} (1/k!)\lambda^{-k/2}(X\cdot\nabla_x + D\cdot\nabla_\xi)^k q(w;0,0)\pi_o(X,D), \tag{4.81}$$

where

$$\sigma_Q(w,\lambda)(x,\xi) = \lambda^m q(w;\lambda^{-\frac{1}{2}}x,\lambda^{-\frac{1}{2}}\xi), \quad (\lambda > 0). \tag{4.82}$$

Consequently, we can obtain the symbol of $S_0QS_0 = T_Q$, since $T_Q u(w) = T_Q(w)u(w)$ with

$$T_Q(y) \sim \sum_{\gamma\geq 0} S_0^{[\gamma]}(QS_0)_{[\gamma]}. \tag{4.83}$$

Note that, after the principal symbol, which is

$$\lambda^m\, q(w;0,0)\pi_o(X,D), \tag{4.84}$$

the next highest order symbol (defining an element of $OP\tilde{\Psi}^{2m-1,\infty}$) is

(4.85) $$\lambda^{m-\frac{1}{2}}\pi_o(X,D)(X\cdot\nabla_x+D\cdot\nabla_\xi)q(w;0,0)\pi_o(X,D)$$

(4.86) $$+\lambda^{m-\frac{1}{2}}\sum_{|\gamma_1|+|\gamma_2|=1}\pi_o^{[0,\gamma_1,\gamma_2]}(X,D)\, q_{[0,\gamma_1,\gamma_2]}(w;0,0)\pi_o(X,D).$$

Now (4.85) always vanishes since X_j and D_j map even functions to odd functions. As for (4.86), if the principal symbol of Q vanishes everywhere on Λ_+, then $q(w;0,0) = 0$ on $\mathbb{H}^n$ and its w-derivatives $q_{[0,\gamma_1,\gamma_2]}(w;0,0)$ also vanish on $\mathbb{H}^n$. Hence T_Q belongs to $OP\tilde{\Psi}^{2m-2,\infty}$ in this case, which is consistent with Theorem A.

Appendix A: The Weyl Calculus

In our development of harmonic analysis on the Heisenberg group, we made considerable use of the Weyl calculus, which is defined as follows. Given $a(x,\xi)$, we set

$$(A.1)\qquad a(X,D) = \iint \tilde{a}(q,p)e^{i(q\cdot X+p\cdot D)}\,dq\,dp$$

where $e^{i(q\cdot X+p\cdot D)}$ is given by (1.5)-(1.7) of Chapter II and $\tilde{a}(q,p)$ is the inverse Fourier transform of $a(x,\xi)$. A few manipulations of integrals give

$$(A.2)\qquad a(X,D)u(x) = (2\pi)^{-n} \int a(\tfrac{1}{2}(x+y),\xi)e^{i(x-y)\cdot\xi}u(y)\,dy\,d\xi.$$

This defines a kernel in $\mathcal{S}'(\mathbb{R}^n \times \mathbb{R}^n)$ and hence a map $a(X,D)\colon \mathcal{S}(\mathbb{R}^n) \to \mathcal{S}'(\mathbb{R}^n)$, for any $a \in \mathcal{S}'(\mathbb{R}^{2n})$. We will touch on only a few properties of the Weyl calculus here, referring to the papers [H10],[G11] for a complete treatment.

One important class of symbols $a(x,\xi)$ is $\mathcal{S}^m_\rho$, defined to consist of smooth $a(x,\xi)$ such that

$$(A.3)\qquad |D^\beta_x D^\alpha_\xi a(x,\xi)| \le C_{\alpha\beta}(1+|x|+|\xi|)^{m-\rho(|\alpha|+|\beta|)}\ .$$

If $a(x,\xi)$ satisfies (A.3), we say $a(X,D) \in OP\mathcal{S}^m_\rho$; generally $\rho \in [0,1]$. If $\rho > 0$, we have the following: $a_j(X,D) \in OP\mathcal{S}^{m_j}_\rho$ implies $a_1(X,D)a_2(X,D) \in OP\mathcal{S}^{m_1+m_2}_\rho$, and the symbol $b(x,\xi)$ of this product has the asymptotic expansion

$$(A.4)\qquad b(x,\xi) \sim \sum_{j\ge 0} (1/j!)\ \{a_1,a_2\}_j(x,\xi)$$

where

$$(A.5)\quad \{a_1,a_2\}_j(x,\xi) = (-i/2)^j \sum_{k=1}^{n} (\partial^2/\partial y_k\partial\xi_k - \partial^2/\partial x_k\partial\eta_k)^j a_1(x,\xi)a_2(y,\eta)\Big|_{\substack{y=x\\ \eta=\xi}}\ .$$

The meaning of (A.4) is that the difference between $b(x,\xi)$ and the sum of the right side of (A.4) over $0 \le j \le N$ belongs to $\mathcal{S}^{-2N\rho+m_1+m_2}_\rho$. We refer to [H10] for a proof of results containing (A.4).

Note that, if $a_1(x,\xi)$ or $a_2(x,\xi)$ is a polynomial in (x,ξ), of degree d, then all the terms in (A.5) are zero for $j > d$. In fact, in this case, (A.4) is not merely asymptotic; we have the identity

$$b(x,\xi) = \sum_{j=0}^{d} (1/j!)\{a_1,a_2\}_j(x,\xi) \tag{A.6}$$

if either $a_1(x,\xi)$ or $a_2(x,\xi)$ is a polynomial of degree d in (x,ξ). The proof of (A.6) is elementary, since one of the factors $a_j(X,D)$ is a differential operator. In particular, if either $a_1(x,\xi)$ or $a_2(x,\xi)$ is a quadratic polynomial, we have

$$b(x,\xi) = a_1(x,\xi)a_2(x,\xi) - \tfrac{1}{2}i\{a_1,a_2\}(x,\xi) - \frac{1}{8}\sum_k(\partial^2/\partial y_k\partial\xi_k - \partial^2/\partial x_k\partial\eta_k)^2 \cdot a_1(x,\xi)a_2(y,\eta)\Big|_{\substack{y=x\\ \eta=\xi}} . \tag{A.7}$$

If we compare the symbol of $a_2(X,D)a_1(X,D)$, we see that only the middle term changes (by a sign), and so, whenever $Q(x,\xi)$ is a quadratic polynomial, the commutator $[Q(X,D),a(X,D)] = C(X,D)$ has symbol

$$C(x,\xi) = -i\{Q,a\}(x,\xi). \tag{A.8}$$

We can use (A.8) to analyze the conjugate of an operator $a(X,D)$ by $e^{iQ(X,D)}$, with $Q(x,\xi)$ quadratic. We claim

$$e^{-isQ(X,D)}a(X,D)e^{isQ(X,D)} = (a\circ g_s)(X,D) \tag{A.9}$$

where $g_s \in Sp(n,\mathbb{R})$ is the one parameter group of symplectic linear maps on $\mathbb{R}^n$ generated by the Hamilton vector field

$$H_Q = \Sigma(\partial Q/\partial x_j\,\partial/\partial\xi_j - \partial Q/\partial\xi_j\,\partial/\partial x_j), \tag{A.10}$$

i.e.,

$$g_s = \exp(sH_Q). \tag{A.11}$$

To prove (A.9), we note that it is equivalent to the operator differential equation

$$\text{(A.12)} \qquad d/ds\ (a\circ g_s)(X,D) = -i[Q(X,D), a\circ g_s(S,D)].$$

Now the left side of (A.12) is clearly $\{Q, a\circ g_s\}(X,D)$, so (A.12) is equivalent to

$$\text{(A.13)} \qquad \{Q,a\}(X,D) = -i[Q(X,D), a(X,D].$$

But this follows from (A.8), so we have proved (A.9).

The set of quadratic polynomials $Q(x,\xi)$, with the Poisson bracket, is naturally isomorphic to the Lie algebra $sp(n,\mathbb{R})$ of the symplectic group $Sp(n,\mathbb{R})$. The map $Q \mapsto iQ(X,D)$ gives a Lie algebra representation of $sp(n,\mathbb{R})$ by skew adjoint operators. This generates a representation of the universal cover of $Sp(n,\mathbb{R})$:

$$\text{(A.14)} \qquad \omega\colon \widetilde{Sp(n,\mathbb{R})} \to \mathcal{U}(L^2(\mathbb{R}^n)).$$

We deduce from (A.9) that, if $j\colon \widetilde{Sp(n,\mathbb{R})} \to Sp(n,\mathbb{R})$ is the natural projection, then

$$\text{(A.15)} \qquad (a\circ j(g))(X,D) = \omega(g)^{-1} a(X,D)\omega(g).$$

As a matter of fact, one can reduce (A.14) to a representation of $Mp(n,\mathbb{R})$, the 2-fold cover of $Sp(n,\mathbb{R})$, called the metaplectic group. We will not go into this; see, e.g., Chapter X of [T5].

We now want to use the identity (A.7) to derive a formula for the Weyl symbol of

$$\text{(A.16)} \qquad e^{-tH} = h_t(X,D), \qquad H = -\Delta + |x|^2 = Q(X,D),$$

where

$$\text{(A.17)} \qquad Q(x,\xi) = |x|^2 + |\xi|^2.$$

We will show that

(A.18) $$h_t(x,\xi) = (\cosh t)^{-n} e^{-(\tanh t)(|x|^2+|\xi|^2)},$$

thus providing an alternate proof of Proposition 3.1 of Chapter II. First, we can see by symmetry that $h_t(x,\xi)$ is a function of $|x|^2+|\xi|^2$. Indeed, the unitary group $U(n)$ is naturally contained in $Sp(n,\mathbb{R})$, and it acts transitively on spheres, so this is a consequence of (A.15), since certainly $\omega(g)$ for $j(g) \in U(n)$ commutes with all functions of H. Thus, if we set

(A.19) $$b_t(X,D) = H\, h_t(X,D),$$

then (A.7) implies

(A.20) $$b_t(x,\xi) = Q(x,\xi)h_t(x,\xi) - \tfrac{1}{4}\sum_k (\partial^2/\partial x_k^2+\partial^2/\partial\xi_k^2)h_t(x,\xi).$$

Note that the Poisson bracket $\{Q,h_t\}$ vanishes in this case. Hence, $h_t(x,\xi)$ must satisfy the equation

(A.21) $$(\partial/\partial t)h_t(x,\xi) = -(|x|^2+|\xi|^2)h_t(x,\xi) + \tfrac{1}{4}\sum_k (\partial^2/\partial x_k^2+\partial^2/\partial\xi_k^2)h_t(x,\xi).$$

If we write

(A.22) $$h_t(x,\xi) = g(t,Q), \quad Q = |x|^2+|\xi|^2,$$

then (A.21) becomes

(A.23) $$\partial g/\partial t = -Qg + Q\,\partial^2 g/\partial Q^2 + n\,\partial g/\partial Q.$$

Now (having peeked at (A.18)) we will make the "inspired guess" that

(A.24) $$h_t(x,\xi) = a(t)e^{b(t)(|x|^2+|\xi|^2)}, \quad \text{i.e., } g(t,Q) = a(t)e^{b(t)Q}.$$

Then the left side of (A.23) is $(a'/a+b'Q)g$ and the right side is $(-Q+Qb^2+nb)g$, so the identity (A.23) is equivalent to

(A.25) $$a'(t)/a(t) = nb(t) \quad \text{and} \quad b'(t) = 1 - b(t)^2.$$

We can solve the second equation for $b(t)$ by separation of variables. Since $h_0(x,\xi) = 1$, we need $b(0) = 0$, so the unique solution is easily seen to be

$$b(t) = -\tanh t.$$

Then the equation $a'/a = -n \tanh t$, with $a(0) = 1$, immediately gives

$$a(t) = (\cosh t)^{-n}.$$

This completes the proof of the identity (A.18).

We can also analyze the Weyl symbol $h_t^Q(x,\xi)$ of

$$e^{-tQ(X,D)} = h_t^Q(X,D) \tag{A.26}$$

for a general positive definite quadratic form $Q(x,\xi)$. If

$$Q(x,\xi) = \Sigma\, \mu_j(x_j^2+\xi_j^2), \tag{A.27}$$

then the proof of (A.18) also gives

$$h_t^Q(x,\xi) = \prod_{j=1}^{n} (\cosh t\, \mu_j)^{-1} \exp -[\Sigma(\tanh t\, \mu_j)(x_j^2+\xi_j^2)]. \tag{A.28}$$

We can express this in invariant form, as follows. Use the Hamilton map F_Q associated with Q, by

$$\sigma(u,F_Q v) = Q(u,v) \tag{A.29}$$

where σ is the symplectic form on $\mathbb{R}^{2n}$ and $Q(u,v)$ is the symmetric bilinear form associated with the quadratic form $Q(u)$ $(u = (x,\xi))$. For Q positive definite, a basic result of symplectic algebra states that F_Q has spectrum $\{\pm i\mu_j\}$, $1 \le j \le n$, $\mu_j > 0$; see, e.g., Chapter I, section 6 of [T5]. Thus

$$\det \cosh i^{-1}tF_Q = (\prod_{j=1}^{n} \cosh t\, \mu_j)^2.$$

Also, if we set

$$A_Q = (-F_Q^2)^{1/2}, \tag{A.30}$$

the unique square root of $-F_Q^2$ with positive spectrum, we obtain (with $\zeta = (x,\xi)$)

$$\Sigma(\tanh t\,\mu_j)(x_j^2+\xi_j^2) = Q(A_Q^{-1}\tanh tA_Q\zeta,\zeta),$$

and hence

$$h_t^Q(x,\xi) = [\det\cosh i^{-1}tF_Q]^{-\frac{1}{2}}\exp[-Q(A_Q^{-1}\tanh tA_Q\zeta,\zeta)]. \tag{A.31}$$

Since any positive definite form $Q(x,\xi)$ can be put in the form (A.27) via a symplectic change of coordinates, the identity (A.15) implies that (A.31) is valid for a general positive definite quadratic form $Q(x,\xi)$, as the formula for the Weyl symbol of $e^{-tQ(X,D)}$.

We turn to a study of L^2 boundedness of operators $a(X,D)$. We will show that $a(X,D)$ is bounded on $L^2(\mathbb{R}^n)$ if $a(x,\xi) \in \mathcal{S}_0^0$; this is a result of Calderon and Vaillancourt [C1]; see also Hörmander [H10]. The approach we take is due to Cordes [C6] and Kato [K2], in a slightly different context. A related approach is given in Howe [H11]. The proof starts with the following two simple operator theoretic results, whose proofs can be found in Chapter 13 of [T2], or the reader can try them as exercises.

Lemma A.1. Let Y be any σ-finite measure space and $U(y)$ a weakly measurable family of bounded operators on a Hilbert space H such that

$$\int_Y |(U(y)f,g)|^2\,dy \le C\,\|f\|^2\,\|g\|^2,\quad f,g \in H. \tag{A.32}$$

If $b \in L^\infty(Y)$, then for any trace class operator G on H,

$$B = b\{G\} = \int_Y b(y)U(y)^*CU(y)\,dy \tag{A.33}$$

satisfies

$$\|B\| \le C\|b\|_{L^\infty}\|G\|_{tr}. \tag{A.34}$$

In our case, we take $Y = \mathbb{R}^{2n}$ with Lebesgue measure, $y = (x,\xi)$. We let $H = L^2(\mathbb{R}^n)$, $X_ju(x) = x_ju(x)$, $D_ju(x) = i^{-1}\partial u/\partial x_j$, and

$$U(y) = U(x,\xi) = e^{i\xi\cdot X+ix\cdot D}\quad. \tag{A.35}$$

Lemma A.2. If $U(y)$ is given by (A.35), then (A.32) is satisfied.

The way in which (A.33), (A.34) will be implemented is described as follows.

Proposition A.3. Let $a = b*g$, $G = g(X,D)$. Then

$$a(X,D) = \iint b(x,\xi)\, e^{i\xi\cdot X - ix\cdot D} G e^{-i\xi\cdot X + ix\cdot D}\, dx\, d\xi. \tag{A.36}$$

Proof. Clearly the superposition principle gives

$$a(X,D) = \int b(x,\xi) g(X-x, D-\xi)\, dx\, d\xi.$$

Thus (A.36) is equivalent to the observation that

$$g(X-x, D-\xi) = e^{i\xi\cdot X - ix\cdot D} g(X,D) e^{-i\xi\cdot X + ix\cdot D}, \tag{A.37}$$

which is elementary (and which complements (A.9)).

We shall apply (A.36) to the case

$$\begin{aligned} b &= (1-\Delta_x-\Delta_\xi)^k a(x,\xi), \\ g(x,\xi) &= (1-\Delta_x-\Delta_\xi)^{-k}\delta, \text{ i.e., } g(q,p) = (1+|q|^2+|p|^2)^{-k}. \end{aligned} \tag{A.38}$$

We will pick $k = k(n)$ sufficiently large below. In such a case, the hypothesis

$$|D^\alpha_{x,\xi} a(x,\xi)| \le A, \quad |\alpha| \le K(n) \tag{A.39}$$

implies $b(x,\xi) \in L^\infty(\mathbb{R}^{2n})$, provided $K(n) = 2k(n)$. It remains to see when $G = g(x,D)$ is trace class. Note that

$$G\, u(x) = \int K(x,y) u(y)\, dy \tag{A.40}$$

where

$$\begin{aligned} K(x,y) &= \int e^{i(x-y)\cdot\xi}\, g(\tfrac{1}{2}(x+y),\xi)\, d\xi \\ &= L(x-y, \tfrac{1}{2}(x+y)). \end{aligned} \tag{A.41}$$

Note that

$$x^\gamma y^\delta D^\alpha_x D^\beta_y L(x,y) = \int e^{ix\cdot\xi} y^\delta \xi^\alpha D^\gamma_\xi D^\beta_y g(y,\xi)\, d\xi. \tag{A.42}$$

Now (A.38) implies $g(x,\xi)$ is C^∞ on $\mathbb{R}^{2n} \setminus 0$ and exponentially decreasing together with all its derivatives, as $|x|+|\xi| \to \infty$. Also g is C^{ℓ} in a neighborhood of the origin, if $k(n) > \ell + n/2$. Thus (A.42) is bounded and continuous for $|\alpha|, |\beta|, |\gamma|, |\delta| \leq \mu$ if $\ell \geq 2\mu$. On the other hand, it is easy to verify that, for $\mu = \mu(n)$ large enough, such a condition implies K is the kernel of a trace class operator. Thus for $G = g(X,D)$ defined by (A.38) with $k = k(n)$ sufficiently large, G is of trace class.

Now, the first two lemmas and proposition A.3 yield the following Calderon-Vaillancourt theorem, which was stated in Chapter II, section 1; see (1.26), (1.27).

Theorem A.4. If the estimate (A.39) holds, K(n) sufficiently large, then $a(X,D): L^2(\mathbb{R}^n) \to L^2(\mathbb{R}^n)$, with

$$\|a(X,D)\| \leq C(n)A. \tag{A.43}$$

In particular, A is bounded on L^2 if $A \in OP\mathcal{S}^0_0$. The class $\mathcal{S}^0_0$ coincides with the class $S^0_{0,0}$, where $a(x,\xi)$ belongs to $S^m_{\rho,\delta}$ if and only if

$$|D^\beta_x D^\alpha_\xi a(x,\xi)| \leq C_{\alpha\beta}(1+|\xi|)^{m-\rho|\alpha|+\delta|\beta|} . \tag{A.44}$$

A more general version of the Calderon-Vaillancourt theorem is that $a(X,D)$ is continuous on $L^2(\mathbb{R}^n)$ provided that, for some $\rho \in [0,1)$, $a(x,\xi) \in S^0_{\rho,\rho}$. This result is particularly important in the case $\rho = \frac{1}{2}$. We will not give the proof of this, but note it can be deduced from Theorem A.4 via the Cotlar-Stein lemma; see Beals [B2] for this argument in the context of the Kohn-Nirenberg calculus (and involving more general classes of symbols).

We now show how the identity (A.18) for the Weyl symbol of e^{-tH}, $H = -\Delta+|x|^2$, helps one to prove sharp Gårding inequalities. That one could do this was pointed out by Unterberger [U1], which concerned itself not with sharp Gårding inequalities, but rather with further generalizations of the Calderon-Vaillancourt theorem. We seek conditions under which a symbol $p(x,\xi) \geq 0$ defines an operator which, if not positive, is at least semi-bounded:

$$\text{(A.45)} \qquad (p(X,D)u,u) \ge -K\,\|u\|^2_{L^2}\ .$$

Now what the identity (A.18) implies is that

$$\text{(A.46)} \quad \alpha_t(x,\xi) = (t/\pi)^{n/2}\, e^{-t(|x|^2+|\xi|^2)} \Rightarrow \alpha_t(X,D) \ge 0 \text{ provided } 0 < t \le 1,$$

since clearly e^{-sH} is a positive operator for any $s \ge 0$. Now operators can be synthesized from $\alpha_t(X,D)$ and various other operators unitarily equivalent to it, and these will provide a sufficiently rich class of positive operators to enable us to obtain sharp Gårding inequalities. One result is the following.

Proposition A.5. The semiboundedness (A.45) holds for all $p(x,\xi)$ such that

$$\text{(A.47)} \qquad p \in \mathcal{S}^2_1.$$

Proof. Consider the operator $p_1(X,D)$ with $p_1 = p*\alpha_1$. Then, as in (A.36),

$$\text{(A.48)} \quad p_1(X,D) = \int p(x,\xi)e^{i\xi\cdot X-ix\cdot D}\alpha_1(X,D)\, e^{-i\xi\cdot X+ix\cdot D}\, dx\, d\xi \ge 0.$$

Consequently (A.45) follows once we establish that

$$\text{(A.49)} \qquad p_1 - p \in \mathcal{S}^0_1,$$

in view of Theorem A.4.

We shall establish (A.49) by a procedure which will also figure in the next proposition. We start with the following characterization of $\mathcal{S}^m_1$. For a given $(x_j,\xi_j) \in \mathbb{R}^{2n}$, let $M_j = \max\{|x_{j1}|,\dots|x_{jn}|,|\xi_{j1}|,\dots,|\xi_{jn}|\}$, where $x_j = (x_{j1},\dots,x_{jn})$, etc., and let Q_j be the 2n-dimensional "cube" of edge M_j, centered at (x_j,ξ_j). (So a concentric cube twice the linear size of Q_j barely contains 0.) Let Q_o be the unit cube centered at the origin in $\mathbb{R}^{2n}$. Let Φ_j denote the natural affine map of Q_o onto Q_j, taking 0 to (x_j,ξ_j). Suppose enough cubes Q_j are picked to cover $\mathbb{R}^{2n}$. Then $q(x,\xi) \in C^\infty(\mathbb{R}^{2n})$ defines a sequence of functions

(A.50) $$q_j(x,\xi) = q \circ \Phi_j \in C^\infty(Q_o).$$

Now q belongs to $\mathcal{S}_1^m$ if and only if it is smooth on $\mathbb{R}^{2n}$ and

(A.51) $$(M_j)^{-m}\, q_j \quad \text{is bounded in } C^\infty(Q_o).$$

We note that if we pick (x_j,ξ_j) appropriately, we can pick a partition of unity $\psi_j(x,\xi) \geq 0$ subordinate to Q_j, bounded in $\mathcal{S}_1^0$. Given $p \in \mathcal{S}_1^m$, then

(A.52) $$(M_j)^{-m}(\psi_j p)\circ\Phi_j \quad \text{is bounded in } C_o^\infty(Q_o).$$

Now, with $p \in \mathcal{S}_1^2$, write $p = \Sigma\psi_j p = \Sigma p_j$, with ψ_j such a partition of unity, so

$$q = p*\alpha_1 - p = \sum_j (p_j*\alpha_1 - p_j).$$

It follows that

$$q\circ\Phi_j = \sum_j (\tilde{p}_j*\alpha_{M_j^2} - \tilde{p}_j)$$

where $\tilde{p}_j = p_j\circ\Phi_j$. On the other hand, it is easy to see that, for $\varphi \in C_o^\infty(Q_o)$,

(A.53) $$\varphi*\alpha_t - \varphi = O(t^{-1}) \quad \text{in } \mathcal{S}(\mathbb{R}^{2n}) \quad \text{as } t \to \infty.$$

Furthermore, for t large, $\varphi*\alpha_t$ only spills outside Q_o to a small degree, so it follows without difficulty that $q\circ\Phi_j$ is bounded in $C^\infty(Q_o)$ if $p_j\circ\Phi_j$ is $O(M_j^2)$ in $C_o^\infty(Q_o)$, so $q \in \mathcal{S}_1^0$, as asserted. This completes the proof of proposition A.5.

A modification of the proof of proposition A.5 yields the following sharp Gårding inequality of Hörmander [H5] and Lax-Nirenberg [L3]:

<u>Proposition A.6.</u> The semiboundedness (A.45) holds for all $p(x,\xi) \geq 0$ such that

(A.54) $$p \in S^1_{1,0}.$$

Proof. This time, we cover $\mathbb{R}^{2n}$ with rectangles R_j centered at (x_j,ξ_j), such that $|x-x_j| \leq \frac{1}{2}$ and $|\xi-\xi_j| \leq \frac{1}{2}M_j$ defines R_j; here $M_j = \max\{|\xi_{j1}|,\dots,|\xi_{jn}|\}$, where $\xi_j = (\xi_{j1},\dots,\xi_{jn})$. Let Ψ_j be the natural affine map from the unit cube Q_o in $\mathbb{R}^{2n}$ onto R_j. Then $q \in C^\infty(\mathbb{R}^{2n})$ defines a sequence of functions $q\circ\Psi_j \in C^\infty(Q_o)$, and q belongs to $S^m_{1,0}$ if and only if $\{(M_j)^{-m}\, q\circ\Psi_j\}$ is bounded in $C^\infty(Q_o)$. Again, it is possible to pick a cover R_j and a subordinate partition of unity $\psi_j(x,\xi) \geq 0$, bounded in $S^0_{1,0}$. Let

$$p_j = (\psi_j p)\circ\Psi_j.$$

Then $p \in S^1_{1,0}$ implies $M_j^{-1}p_j$ bounded in $C_o^\infty(Q_o)$. Now define $\tilde{p}_j \in \mathcal{S}(\mathbb{R}^{2n})$ by

$$\tilde{p}_j\circ\Psi_j = p_j * \alpha_{M_j} .$$

This means that

$$\tilde{p}_j = (\psi_j p) * \beta_j$$

where β_j is obtained from α_1 via a linear symplectic transformation. It follows that $\beta_j(X,D) \geq 0$ and, since $\psi_j(x,\xi)p(x,\xi) \geq 0$, we have

$$\tilde{p}_j(X,D) \geq 0. \tag{A.55}$$

Now let

$$\tilde{p}(x,\xi) = \sum_j \tilde{p}_j(x,\xi).$$

Then (A.55) implies

$$\tilde{p}(X,D) \geq 0. \tag{A.56}$$

On the other hand, the same sort of argument used in proposition A.5 shows

$$\tilde{p} - p \in S^0_{1,0} , \tag{A.57}$$

so $\tilde{p}(X,D) - p(X,D)$ is a bounded operator on L^2. This proves proposition A.6.

Proposition A.5 has the appearance of a weak form of the Fefferman-Phong inequality. In fact, Fefferman and Phong [F1] prove that the semiboundedness (A.45) holds for all $p(x,\xi) \geq 0$ such that $p \in S^2_{1,0}$, if p is scalar. This result clearly is more powerful than either proposition A.5 or A.6 and contains them, in the scalar case. On the other hand, these weaker "sharp Gårding inequalities" have proofs that hold for systems, whereas the Fefferman-Phong inequality has not been established for systems.

Appendix B: Toeplitz Operators and More General Weyl Calculus

A principal goal of this appendix is to discuss the following result of R. Howe: there is a natural unitary map from $L^2(\mathbb{R}^n)$ to $L^2_{\mathcal{H}}(B^n)$, the space of holomorphic functions L^2 on the unit ball $B^n \subset \mathbb{C}^n$, which sets up an isomorphism between the algebra $OP\mathcal{H}^m_b$ of pseudodifferential operators on $\mathbb{R}^n$ and the algebra of Toeplitz operators on B^n of order $m/2$. This elegant result was given in [H11] and also discussed in [G15]. Since these references confine the proof to a sketch, we shall take the space to provide a full proof here. The proof will arise as an application of a more general sort of Weyl calculus than that dealt with in Appendix A.

The unitary isomorphism of $L^2(\mathbb{R}^n)$ with $L^2_{\mathcal{H}}(B^n)$ starts with the unitary transformation intertwining the representation π_1 of the Heisenberg group $\mathbb{H}^n$ on $L^2(\mathbb{R}^n)$ (given in Chapter II, section 1) with the Bargman-Fok representation β of $\mathbb{H}^n$, which we will now define. The representation β acts on the Hilbert space

$$\mathcal{H} = \{u(z) \text{ holomorphic on } \mathbb{C}^n: \int |u(z)|^2 e^{-|z|^2/2}\, d\,\mathrm{vol}(z) < \infty\}. \tag{B.1}$$

Then, for $(t,q,p) \in \mathbb{H}^n$, the unitary operator $\beta(t,q,p)$ on $\mathcal{H}$ is defined by

$$\beta(t,q,p)u(z) = e^{it+i(q+ip)\cdot z\sqrt{2}-(|q|^2+|p|^2)}u(z+i(q-ip)/\sqrt{2}). \tag{B.2}$$

Note that, on the Lie algebra level,

$$\beta(T) = iI,\ \beta(L_j) = (\partial/\partial z_j+z_j)\sqrt{2},\ \beta(M_j) = (\partial/\partial z_j-z_j)/i\sqrt{2}. \tag{B.3}$$

Recall that, for the Schrodinger representation of $\mathbb{H}^n$ on $L^2(\mathbb{R}^n)$,

$$\pi_1(T) = iI,\ \pi_1(L_j) = \partial/\partial x_j,\ \pi_1(M_j) = ix_j. \tag{B.4}$$

The unitary operator $W: L^2(\mathbb{R}^n) \to \mathcal{H}$ intertwining π_1 and β is given by

(B.5) $\quad Wf(z) = \int f(x)K(x,z)\,dx, \quad K(x,z) = \exp[\sqrt{2}\, z\cdot x - (z\cdot z+|x|^2)/2].$

Now there is a natural unitary isomorphism

(B.6) $$V\colon \mathcal{H} \to L^2(B^n),$$

the space of L^2 functions holomorphic on the unit ball $B^n \subset \mathbb{C}^n$, which arises as follows. An orthonormal basis on $\mathcal{H}$ is

(B.7) $$u_\alpha = a_\alpha z^\alpha, \quad a_\alpha = \sqrt{2/\alpha!}\,,$$

(see, e.g., [T5], Chapter I), while an orthonormal basis of $L^2_{\mathcal{H}}(B^n)$ is given by

(B.8) $$v_\alpha = b_\alpha z^\alpha, \quad b_\alpha = \sqrt{(n+|\alpha|)!/\alpha!}\quad .$$

See Rudin [R7]. Thus, we define V by

(B.9) $\quad Vu_\alpha = v_\alpha$, i.e., $Vz^\alpha = \gamma_\alpha z^\alpha, \ \gamma_\alpha = b_\alpha/a_\alpha = 2^{\frac{1}{2}}\pi^{-n/2}\sqrt{(n+|\alpha|)!} = \gamma_{|\alpha|}.$

Now we want to analyze the operators $\mathcal{Z}_j, \mathcal{L}_j$, defined by

(B.10) $$\mathcal{Z}_j = Vz_jV^{-1}, \ \mathcal{L}_j = V\partial/\partial z_jV^{-1}.$$

In fact, (B.9) shows that

$$\mathcal{Z}_j z^\alpha = (\gamma_{|\alpha|+1}/\gamma_{|\alpha|})z_jz^\alpha, \ \mathcal{L}_j z^\alpha = (\gamma_{|\alpha|-1}/\gamma_{|\alpha|})(\partial/\partial z_j)z^\alpha.$$

In particular,

(B.11) $$\mathcal{Z}_j z^\alpha = \sqrt{|\alpha|+n+1}\ \ z_jz^\alpha.$$

The operator $z^\alpha \mapsto |\alpha|z^\alpha$ can be extended to $C^\infty(B^n)$ as $u(z) \mapsto i^{-1}Xu$ where X is the real vector field on $\mathbb{R}^{2n} = \mathbb{C}^n$ generating the flow $z \mapsto e^{it}z$ on $\mathbb{C}^n$. If we set $D = i^{-1}X$, a first order differential operator, we get

(B.12) $$\mathcal{Z}_j = z_j\sqrt{|D|+n+1}\ .$$

Sinze z_j and $\partial/\partial z_j$ are adjoints of each other on $\mathcal{H}$, the adjoint of $\mathcal{Z}_j$ on $L^2(B^n)$ is $\mathcal{L}_j$. Thus, if ν is the unitary representation of $\mathbb{H}^n$ on $L^2_{\mathcal{H}}(B^n)$ defined by

$$\text{(B.13)} \qquad \nu(g) = V\beta(g)V^{-1}, \qquad g \in \mathbb{H}^n,$$

then

$$\text{(B.14)} \qquad \begin{aligned} \nu(T) &= iI, \ \nu(L_j) = 2^{-3/2}\pi(z_j(|D|+n+1)^{\frac{1}{2}} + (|D|+n+1)^{\frac{1}{2}}\bar{z}_j)\, i, \\ \nu(M_j) &= 2^{-3/2}\pi(z_j(|D|+n+1)^{\frac{1}{2}} - (|D|+n+1)^{\frac{1}{2}}\bar{z}_j), \end{aligned}$$

on $L^2_{\mathcal{H}}(B^n)$. This was derived in [H11] and [G15]. Let us denote by π the orthogonal projection of $H^{-\frac{1}{2}}(S^{2n-1})$ on $L^2_{\mathcal{H}}(B^n)$, which we identify with a closed linear subspace of the Sobolev space $H^{-\frac{1}{2}}(S^{2n-1})$ via the Poisson integral. The projection π is essentially the Szegő projector. We can write

$$\text{(B.15)} \qquad \nu(L_j) = i\pi\mathcal{D}_j\pi, \quad \nu(M_j) = i\pi \ \mathcal{X}_j\pi$$

where

$$\text{(B.16)} \quad \mathcal{D}_j = z_j(|D|+n+1)^{\frac{1}{2}} + (|D|+n+1)^{\frac{1}{2}}\bar{z}_j, \qquad i\mathcal{X}_j = z_j(|D|+n+1)^{\frac{1}{2}} - (|D|+n+1)^{\frac{1}{2}}\bar{z}_j.$$

Now $\mathcal{D}_j$ and $\mathcal{X}_j$ are not quite pseudodifferential operators. But note that, for all $u \in \mathcal{D}'(S^{2n-1})$, πu has wave front set in Λ_+, a ray bundle in $T^*S^{2n-1} \setminus 0$ defining the contact structure, and the differential operator D is noncharacteristic on Λ_+, and so elliptic on a conic neighborhood of Λ_+. Thus, when analyzing $\mathcal{D}_j$ and $\mathcal{X}_j$, we can treat $\mathcal{D}_j$ and $\mathcal{X}_j$ as pseudodifferential operators on S^{2n-1}, belonging to $OPS^{\frac{1}{2}}$. Note that, on Λ_+,

$$\text{(B.17)} \qquad \sigma_{\mathcal{X}_j}(\omega,\xi) = (\operatorname{Re}\omega_j)|\xi|^{\frac{1}{2}}, \ \sigma_{\mathcal{D}_j}(\omega,\xi) = (\operatorname{Im}\omega_j)|\xi|^{\frac{1}{2}} \mod S^{-\frac{1}{2}},$$

where $\omega = (\omega_1,\dots,\omega_n) \in S^{2n-1} \hookrightarrow \mathbb{C}^n$. It is well known that $\mathcal{X}_j$ and $\mathcal{D}_j$ can be replaced by pseudodifferential operators commuting with π, without changing (B.15) or (B.17), and we will assume this done. See the result quoted as Theorem B near the end of Chapter III, section 4. The map

$$U = VW\colon L^2(\mathbb{R}^n) \to L^2_{\mathcal{H}}(B^n)$$

is unitary, and under it, a pseudodifferential operator $a(X,D)$ is taken to

$$U\, a(X,D)\, U^{-1} = \pi a(\mathcal{X},\mathcal{D})\pi, \tag{B.18}$$

where $a(\mathcal{X},\mathcal{D})$ is a function of the self adjoint operators $\mathcal{X}_1,\dots,\mathcal{X}_n,\mathcal{D}_1,\dots,\mathcal{D}_n$ defined by the Weyl calculus:

$$a(\mathcal{X},\mathcal{D}) = \int \tilde{a}(q,p)e^{iq\mathcal{X} + ip\cdot\mathcal{D}}dq\, dp, \tag{B.19}$$

where $q\cdot\mathcal{X} = \Sigma\, q_j\mathcal{X}_j$, $p\cdot\mathcal{D} = \Sigma\, p_j\mathcal{D}_j$.

With this in mind, we will consider the Weyl calculus in the following more general context. On a compact manifold M, take $L_1,\dots,L_k$ self adjoint operators,

$$L_j \in OPS^a, \qquad 0 < a \le 1. \tag{B.20}$$

We suppose L_j have scalar principal symbols. We also suppose

$$\mathcal{L} = L_1^2 + \cdots + L_k^2 \in OPS^{2a} \text{ is elliptic.} \tag{B.21}$$

We want to analyze the operator $a(L)$, defined by the formula

$$a(L) = \int \tilde{a}(s)\; e^{is\cdot L}\; ds, \tag{B.22}$$

as a pseudodifferential operator, given $a(\xi_1,\dots,\xi_n) \in \mathcal{S}^m_\rho$, i.e.,

$$|D^\alpha a(\xi)| \le C_\alpha(1+|\xi|)^{m-\rho|\alpha|}. \tag{B.23}$$

In (B.22), we take $s\cdot L = s_1L_1 + \cdots + s_kL_k$, $s \in \mathbb{R}^k$. Note that (B.23) implies $\tilde{a}(s)$ is C^∞ outside the origin and rapidly decreasing, with all its derivatives, as $|s| \to \infty$. We can localize the analysis of $a(L)$ under the following hypothesis:

$$a \in \mathcal{S}(R^k) \Rightarrow a(L) \in OPS^{-\infty}. \tag{B.24}$$

We will go into cases where (B.24) is sure to be valid shortly. Now, in

(B.22), we can suppose a(s) is supported on a small set $|s| \le \varepsilon$. We will use the method of geometrical optics to obtain a parametrix for $e^{is\cdot L}$, for $|s| \le \varepsilon$. In fact, let $A \in OPS^a$ be some real linear combination of $L_1,\dots,L_k$, and we represent e^{itA}, for $|t|$ small, in the form (in local coordinates)

$$e^{itA}\, u(x) = \int a(t,x,\xi)\, e^{i\varphi(t,x,\xi)+ix\cdot\xi}\hat{u}(\xi)\, d\xi.$$

Here, the phase φ satisfies the eikonal equation

$$\partial\varphi/\partial t = a_1(x,\xi+\nabla_x\varphi), \quad \varphi(0,x,\xi) = 0,$$

where $a_1(x,\ ) \in S^a$ is the principal symbol of $A \in OPS^a_{1,0}$. Thus for small t we can solve for φ, real valued in S^a. We get a sequence of transport equations for $a \sim \Sigma a_j$, yielding $a(t,x,\xi) \in S^0_{1,0}$, with $a(0,x,\xi) = 1$. Writing $A = \Sigma s_j L_j$, the functions φ and a depend smoothly on some set $|s| \le s_0$, $|t| \le t_0$. We can freeze t at t_0 and rescale s, to write, for $|s| \le \varepsilon$,

$$\text{(B.25)} \qquad e^{is\cdot L}\, u(x) = \int b(s,x,\xi)e^{i\psi(s,x,\xi)+ix\cdot\xi}\hat{u}(\xi)\, d\xi.$$

Here $b(0,x,\xi) = 1$, $b \in S^0$, and $\psi \in S^a_{1,0}$ is real valued and satisfies

$$\text{(B.26)} \qquad (\partial\psi/\partial s_j)(0,x,\xi) = \sigma_{L_j}(x,\xi), \quad \psi(0,x,\xi) = 0.$$

Returning to (B.22), we have, for $a \in \mathcal{S}^m_\rho$ such that $\tilde{a}(s)$ is supported on $|s| \le \varepsilon$,

$$\text{(B.27)} \qquad a(L) = \iint \tilde{a}(s)\, b(s,x,\xi)e^{i\psi(s,x,\xi)}e^{ix\cdot\xi}\hat{u}(\xi)\, d\xi\, ds$$
$$= \int a(D_s)(be^{i\psi})\Big|_{s=0}\, e^{ix\cdot\xi}\hat{u}(\xi)\, d\xi.$$

Now we can use the stationary phase method to evaluate $a(D_s)(be^{i\psi})$ provided we know $\nabla_s\psi \ne 0$ at $s = 0$. In view of (B.26), this is equivalent to the hypothesis (B.21). Thus we obtain, for $a \in \mathcal{S}^m_\rho$, provided $\frac{1}{2} < \rho \le 1$,

$$\text{(B.28)} \qquad a(D_s)(be^{i\psi}) = B\, e^{i\psi}$$

where

(B.29) $\quad B \in S^{ma}_{\rho',1-\rho'}$

has an asymptotic expansion

(B.30) $\quad B(s,x,\xi) \sim b(s,x,\xi)a(\nabla_s\psi) + \cdots.$

In particular,

$$B(0,x,\xi) \sim a(\sigma_{L_1}(x,\xi),\ldots,\sigma_{L_k}(x,\xi)) + \cdots.$$

Now (B.27) gives

(B.31) $\quad a(L)u = \int B(0,x,\xi)e^{ix\cdot\xi}\hat{u}(\xi)\, d\xi,$

and this implies a(L) is a pseudodifferential operator. In particular, for $a \in \mathcal{S}^m_1$, we have $a(L) \in OPS^{ma}_{1,0}$, and

(B.32) $\quad \sigma_{a(L)}(x,\xi) = a(L_1(x,\xi),\ldots,L_k(x,\xi)) \in \text{mod } S^{ma-a}_{1,0}$

where $L_j(x,\xi)$ is the principal symbol of L_j.

In general, one might not have (B.24), but the operator calculus (B.31)-(B.32) still holds for $a \in \mathcal{S}^m_\rho$ such that $\tilde{a}(s)$ is supported in a small neighborhood of 0. This can still be quite useful, as we will see below. One case where we can guarantee (B.24) is when $L_1,\ldots,L_k$ all commute. In that case, one has

$$L_j^2\, a(L) = \int D^2_{s_j} a(s)\, e^{is\cdot L}\, ds$$

and hence, if $a \in \mathcal{S}(\mathbb{R}^k)$,

$$\mathcal{L}^j\, a(L) = \int \Delta^j\tilde{a}(s)\, e^{is\cdot L}\, ds$$

is bounded on each Sobolev space $H^s(M)$ for all j. Since $\mathcal{L}$ is elliptic, this implies $a(L) \in OPS^{-\infty}$, if $a \in \mathcal{S}(\mathbb{R}^k)$. In this commutative case, the analysis of the functional calculus given here is equivalent to that given by the author in [T3], and in Chapter XII of [T2]. Other approaches to functional calculi in the commutative case were given by Strichartz [S9] and Colin de Verdiere [C5].

This argument can be pushed to yield (B.24) as long as

$$D_{s_j} e^{is\cdot L} = B_j(s)\, e^{is\cdot L} + L_j\, e^{is\cdot L} \tag{B.33}$$

where each $B_j(s)$ is a polynomial in s with values in $OPS^{a-\varepsilon}$, for some $\varepsilon > 0$. For example, say $L_j \in OPS^{\frac{1}{2}}$ and $B_j(s)$ are polynomials with coefficients in OPS^0. We leave the straightforward verification of this to the reader. Since $\pi\mathfrak{X}_j\pi, \pi\mathfrak{D}_j\pi$ satisfy the same commutation relations as X,D, we see that such an identity holds in this case, so

$$a \in \mathcal{S} \Rightarrow \pi a(\mathfrak{X},\mathfrak{D})\pi \text{ is smoothing.} \tag{B.34}$$

Thus the geometrical optics analysis applies, to give

$$\pi a(\mathfrak{X},\mathfrak{D}) = \pi P_a \pi \tag{B.35}$$

where

$$a \in \mathcal{S}_1^m \Rightarrow P_a \in OPS^{m/2}_{1,0}(S^{2n-1}). \tag{B.36}$$

Suppose $a \in \mathcal{S}^m$, i.e., $a(x,\xi) \sim \Sigma a_j(x,\xi)$ with a_j homogeneous of degree m-j in (x,ξ). In this case, since $\mathfrak{X}_j,\mathfrak{D}_j \in OPS^{\frac{1}{2}}$ (microlocally near Λ_+), we see that the phase function $\psi(s,x,\xi)$, obtained as a solution to an eikonal equation, is asymptotic to $\psi_1+\psi_2+\cdots$, with ψ_{1+k} homogeneous of degree $1-\frac{1}{2}k$. Also the amplitude $B(s,x,\xi)$ is asymptotic to $b_0+b_1+b_2+\cdots$ with b_j homogeneous of degree $-\frac{1}{2}j$ in ξ. It follows that (B.35) holds with

$$a \in \mathcal{S}^m \Rightarrow P_a \sim P_1+P_2+\cdots, \quad P_j \in OPS^{\frac{1}{2}m-\frac{1}{2}j}. \tag{B.37}$$

Now, if $a(x,\xi)$ is equal to a function homogeneous of degree m for $|x|+|\xi|$ large, the expansion above P_a can be improved. In fact, if we use our class of symbols

$$\mathcal{H}^m_b = \{a(x,\xi) : a \sim \Sigma a_j,\ a_j \text{ homogeneous of degree } m-2j\} \tag{B.38}$$

then we will prove the following.

Proposition B.1. If $a(x,\xi) \in \mathcal{H}_b^m$, then (B.35) holds with $P_a \in OPS^{m/2}$, so

(B.39) $$P_a \sim P_1+P_2+\cdots, \quad P_j \in OPS^{\frac{1}{2}m-j},$$

The content of this proposition is that for $a \in \mathcal{H}_b^m$, the terms in the expansion (B.37) vanish for j even. The amplitude $B(0,x,\xi)$ in (B.31) arises by an inductive construction, and such a construction does not lend itself easily to proving this infinite sequence of identities. We will take a different approach, one involving the Weyl functional calculus in a context not quite covered by hypothesis (B.20).

Namely, we need a functional calculus for the quadratic forms in X and D occurring in the metaplectic representation (on the Lie algebra level). Thus, with $sp(n,\mathbb{R}) \approx \mathcal{P}_2$, the linear space of second order homogeneous polynomials $Q(x,\xi)$, we pick a basis Q_j for this vector space and set

(B.40) $$p(Q_1(X,D),\ldots,Q_N(X,D)) = \int \tilde{p}(s)e^{is_1Q_1(X,D)+\cdots+s_NQ_N(X,D)}\, ds$$

where $N = \dim sp(n,\mathbb{R}) = \frac{1}{2}n(n+1)$. To write it in a more invariant fashion, let $\mathbb{Q}$ denote the linear map from $sp(n,\mathbb{R})$ to self adjoint operators on $\mathbb{R}^n$ given by

(B.41) $$\mathbb{Q}(Q) = Q(X,D),$$

and set

(B.42) $$p(\mathbb{Q}) = \int \tilde{p}(Q)\, e^{i\mathbb{Q}(Q)}\, dQ$$
$$= \int \tilde{p}(Q)\, e^{iQ(X,D)}\, dQ.$$

We will parallel the analysis of (B.22), via a representation of $e^{iQ(X,D)}$ as a Fourier integral operator and use of stationary phase, as in (B.27). Our next order of business is hence to represent $e^{iQ(X,D)}$ as a Fourier integral operator for $Q \in \mathcal{P}_2$.

That this can be done is related to the fact that commutators of operators in OPS^{m_j} belong to $OPS^{m_1+m_2-2}$, so operators in OPS^2 behave for many purposes like pseudodifferential operators of order 1 on compact manifolds. The Fourier integral representation of $e^{iQ(X,D)}$ can be derived by the method of geometrical optics (see, e.g., Helffer and Robert [H4]), but we will derive it from the identity (A.31) of Appendix A, which says that, for $Q(x,\xi) \in \mathcal{P}_2$ positive definite,

$$e^{-tQ(X,D)} = h_t^Q(X,D) \tag{B.43}$$

with

$$h_t^Q(x,\xi) = [\det \cosh (t/i)F_Q]^{-\frac{1}{2}} \exp[-Q(A_Q^{-1} \tanh tA_Q\, \zeta,\zeta)]. \tag{B.44}$$

Here $\zeta = (x,\xi)$, $Q(u,v)$ is the bilinear form such that $Q(\zeta,\zeta) = Q(\zeta)$. Recall that F_Q is the linear transformation defined on $\mathbb{R}^{2n}$ by

$$Q(u,v) = \sigma(F_Q u,v) \tag{B.45}$$

and $A_Q = (-F_Q^2)^{\frac{1}{2}}$. Compare (3.55)-(3.60) of Chapter II. If we set

$$\theta(it) = t^{-1} \tanh t, \tag{B.46}$$

which is an even function of t, we can write (B.44) as

$$h_t^Q(x,\xi) = [\det \cosh (t/i)F_Q]^{-\frac{1}{2}}\, e^{-tQ(\theta(tF_Q)\zeta,\zeta)} . \tag{B.47}$$

We can analytically continue to imaginary time, and, with

$$\mathcal{I}(it) = t^{-1} \tan t, \tag{B.48}$$

we get

$$h_{it}^Q(x,\xi) = [\det \cos (t/i)F_Q]^{-\frac{1}{2}}\, e^{itQ(\mathcal{I}(tF_Q)\zeta,\zeta)} . \tag{B.49}$$

Consequently we have the Fourier integral representation

$$e^{iQ(X,D)}\, u(x) = \int a(Q)\, e^{i\varphi(Q,\frac{1}{2}(x+y),\xi)+i(x-y)\cdot\xi} u(y)\, dy\, d\xi \tag{B.50}$$

where

(B.51) $\varphi(Q,x,\xi) = -Q(\mathcal{I}(F_Q)\zeta,\zeta), \quad \zeta = (x,\xi),$

and

(B.52) $a(Q) = [\det \cos F_Q/i]^{-\frac{1}{2}}$

Now (B.50) was derived under the hypothesis that $Q(x,\xi)$ was positive definite, but it is clear that the resulting formula is valid for all $Q \in \mathfrak{P}_2$. The amplitude $a(Q)$ and phase φ have singularities at $Q \in \mathfrak{P}_2$ for which $\frac{1}{i}F_Q$ has $(k+\frac{1}{2})\pi$ as an eigenvalue for some integer k, but formula (B.50) is valid, with $a(Q)$ and φ smooth, for Q in a neighborhood $\mathcal{O}$ of the origin.

If $\tilde{p}$ is supported in $\mathcal{O}$, $p \in \mathcal{S}^m$, then (B.42) and (B.50) give

(B.53) $$p(\mathbb{Q})u(x) = \iint \tilde{p}(Q)(a(Q)\; e^{i\varphi(Q,\frac{1}{2}(x+y),\xi)})e^{i(x-y)\cdot\xi}\; u(y)\; dy\; d\xi\; dQ$$
$$= \int p(D_Q)(a(Q)e^{i\varphi(Q,\frac{1}{2}(x+y),\xi)})\Big|_{Q=0}\; e^{i(x-y)\cdot\xi}u(y)\; dy\; d\xi.$$

Now, in view of (B.51), we see that $\nabla_Q\varphi(0,x,\xi) = Q(x,\xi)$, so the Q-gradient vector of φ is nonvanishing for $(x,\xi) \neq (0,0)$, and hence the stationary phase method yields

(B.54) $$p(\mathbb{Q})u(x) = \int b(0,\tfrac{1}{2}(x+y),\xi)e^{i(x-y)\cdot\xi}\; u(y)\; dy\; d\xi$$
$$= b(0,X,D)\; u(x)$$

where

(B.55) $$b(Q,\tfrac{1}{2}(x+y),\xi) = e^{-i\varphi}p(D_Q)(a(Q)e^{i\varphi}).$$

If $p \in \mathcal{S}^m$, we obtain the asymptotic expansion

(B.56) $$b(Q,x,\xi) \sim \Sigma b_j(Q,x,\xi)$$

with $b_j(Q,x,\xi)$ homogeneous of degree $2(m-j)$. (Note that $b_j(Q,x,\xi) = b_j(Q,-x,-\xi)$.) Thus

(B.57) $$p \in \mathcal{S}^m \Rightarrow p(\mathbb{Q}) \in OP\mathcal{H}_b^{2m},$$

if $\tilde{p}$ is supported sufficiently near the origin, and we have

(B.58) $$\sigma_{p(\mathbb{Q})}(x,\xi) = p(\tilde{Q}(x,\xi)) \mod \mathcal{H}_b^{2m-2}.$$

Here we define $\tilde{Q}: \mathbb{R}^{2n} \to \mathfrak{P}_2' \approx sp(n,\mathbb{R})'$, the linear dual of $\mathfrak{P}_2$, by

(B.59) $\langle \tilde{Q}(x,\xi), Q \rangle = Q(x,\xi).$

Using (B.57) and (B.58), we will establish the following.

Lemma B.2. Let $q(x,\xi) \in \mathcal{H}_b^m$ be even, i.e., $q(x,\xi) = q(-x,-\xi)$. Then there exists $p \in \mathcal{S}^{m/2}$ such that

(B.60) $p(\mathbb{Q}) - q(X,D) \in OP\mathcal{S}_1^{-\infty}.$

Proof. We begin with the observation that if $q(x,\xi) \in \mathcal{H}_b^m$ is even, there is a smooth function $p_o \in \mathcal{S}^{m/2}$ such that $p_o(x_jx_k, x_j\xi_k, \xi_j\xi_k)$ is equal to $q(x,\xi)$, outside some neighborhood of the origin. This follows from the result of A. Schwartz [S2]; see also Mather [M1]. This argument was noted in another context in Guillemin and Sternberg [G15] (see the proof of Lemma 6.14 in [G15]). One can truncate $\tilde{p}_o$ to preserve evenness, so $\tilde{p}_o$ is supported near the origin. This alters p_o by an element of $\mathcal{S}_1^{-\infty}$. For such p_o, (B.57)-(B.58) give

(B.61) $p_o(\mathbb{Q}) - q(X,D) \in OP\mathcal{H}_b^{m-2}.$

Now an inductive argument yields $p_j(\mathbb{Q}) \in OP\mathcal{H}_b^{m-2j}$ such that, if $p \sim \Sigma p_j$, (B.60) holds.

We return to our analysis of

(B.62) $\pi a(\mathfrak{X},\mathfrak{D})\pi = Ua(X,D)U^{-1}$

for $a(x,\xi) \in \mathcal{H}_b^m$. Recall that we have

(B.63) $\pi a(\mathfrak{X},\mathfrak{D})\pi = \pi P_a \pi$

with $P_a \in OPS_{1,0}^{m/2}$ being asymptotic to ΣP_j, $P_j \in OPS^{m-\frac{1}{2}j}$. Now suppose $a(x,\xi) \in \mathcal{H}_b^m$ is even. Then, by Lemma B.2, we can write

(B.64) $a(X,D) = p(\mathcal{Q}) \quad \text{mod } OP\mathcal{S}_1^{-\infty}$,

where $p \in \mathcal{S}^{m/2}$. Now if we define a linear map Σ from $\mathcal{P}_2 \approx sp(n,\mathbb{R})$ to self adjoint operators by

(B.65) $\Sigma(Q) = Q(\mathfrak{X},\mathfrak{D})$,

we have, modulo a smoothing operator,

(B.66) $\pi a(\mathfrak{X},\mathfrak{D})\pi = \pi p(\Sigma)\pi$.

Note that, by (B.16)-(B.17), for each $Q \in \mathcal{P}_2$, $Q(\mathfrak{X},\mathfrak{D}) \in OPS^1$ microlocally near Λ_+, and the functional calculus for $p(L)$, $L_j \in OPS^1$ developed above applies to $p(\Sigma)$. In other words, on a conic neighborhood of $\Lambda_+ \subset T^*(S^{2n-1}) \setminus 0$,

(B.67) $p \in \mathcal{S}^{m/2} \subset p(\Sigma) \in OPS^{m/2}$.

Consequently we have proved:

(B.68) $a(x,\xi) \in \mathcal{H}_b^m$ <u>even</u> $\Rightarrow \pi a(\mathfrak{X},\mathfrak{D})\pi = \pi P_a \pi \quad$ with $P_a \in OPS^{m/2}$.

We are now in a position to prove Proposition B.1. In fact, the weak result (B.37) implies that the correspondence $a(X,D) \mapsto P_a$ is "microlocal," and in analyzing P_a on a microlocal level there is no loss of generality in supposing $a(x,\xi)$ is even. From Proposition B.1 easily follows the more rounded out statement of affairs which we record here:

<u>Theorem B.3</u>. With the unitary map U given above, the correspondence

(B.69) $a(X,D) \mapsto Ua(X,D)U^{-1} = \pi a(\mathfrak{X},\mathfrak{D})\pi$

sets up an isomorphism between $OP\mathcal{H}_b^m$ and $\mathcal{J}^{m/2}$, where

(B.70) $\mathcal{J}^{m/2} = \{\pi P \pi : P \in OPS^{m/2}\}$

is the space of Toeplitz operators of order m/2.

Proof. Here $OPS^{m/2}$ refers to pseudodifferential operators on the sphere S^{2n-1}. We need only check the surjectivity. If $\pi P \pi \in \mathcal{J}^{m/2}$, the construction above produces $a_0(X,D) \in OP\mathcal{H}_b^m$ whose image in $\mathcal{J}^{m/2}$ is $\pi P_1 \pi$, where $P_1 \in OPS^m$ has the same principal symbol as that of P on Λ_+. Now we appeal to Theorem A, stated near the end of section 4, Chapter III. The difference $\pi(P-P_1)\pi$ is a Toeplitz operator of order m/2 - 1. Inductively we obtain $a_j(X,D) \in OP\mathcal{H}_b^{m-2j}$ which asymptotically sum to a(X,D), whose image differs from $\pi P \pi$ by a smoothing operator, which in turn is readily absorbed. The proof is complete.

Note that Theorem B.3 implies an equivalence between index theorems for elliptic operators in $OP\mathcal{H}_b^m$ (contained, e.g., in [H10]),and the index theorem of Venugopalkrishna [V2] for Toeplitz operators on B^n. Boutet de Monvel [B8] has index theorems for general strictly pseudoconvex domains. Also Theorem B.3 implies an equivalence between results on spectral asymptotics for elliptic operators obtained by Helffer and Robert [H4], Shubin [S5], Chazarain [C2], and others, and results on spectral asymptotics for elliptic Toeplitz operators on B^n. Such results on spectral asymptotics, for elliptic Toeplitz operators on general strictly pseudoconvex domains, have been proved by Boutet de Monvel and Guillemin [B11]. Also, Guillemin and Sternberg [G15] analyze spectral asymptotics for $a(X,D) \in OP\mathcal{H}_b^m$, in the special case when $a(x,\xi)$ is even, using a different "compactification."

Index of Symbol Classes

For pseudodifferential operators on $\mathbb{R}^n$, we have an operator $p(x,D)$ associated to a symbol $p(x,\xi)$, via formula (1.1) of Chapter I. In this case, symbol classes of particular importance include $S^m_{\rho,\delta}$, consisting of $p(x,\xi)$ such that $|D^\beta_x D^\alpha_\xi p(x,\xi)| \le C_{\alpha\beta}(1+|\xi|)^{m-\rho|\alpha|+\delta|\beta|}$, and the subclass S^m of $S^m_{1,0}$, consisting of $p(x,\xi)$ asymptotic to

$$p(x,\xi) \sim \sum_j p_{m-j}(x,\xi)$$

where, for $|\xi| \ge 1$, $p_{m-j}(x,\xi)$ is homogeneous of degree m-j in ξ. A large number of other classes of "symbols" has arisen in this paper, to which operators have been associated, by various rules. We provide a list of these classes here for convenient reference.

Index of Other Notation

The following is a list of other special symbols and the page where they are introduced.

References

[A1]. R. Anderson, "The Weyl functional calculus," J. Funct. Anal. 4(1969), 240-267.

[B1]. M. Beals, C. Fefferman, and R. Grossman, "Strictly pseudoconvex domains in C^n," Bull. AMS 8(1983), 125-322.

[B2]. R. Beals, "A general calculus of pseudodifferential operators," Duke Math. J. 42(1979), 1-42.

[B3]. _____, "Spatially inhomogeneous pseudodifferential operators II," Comm. Pure Appl. Math. 27(1974), 161-205.

[B4]. _____, "Weighted distribution spaces and pseudodifferential operators," J. d'Anal. Math. 39(1981), 131-187.

[B5]. R. Beals and C. Fefferman, "Spatially inhomogeneous pseudodifferential operators," Comm. Pure Appl. Math. 27(1974), 1-24.

[B6]. R. Beals and P. Greiner, "Pseudodifferential operators associated to hyperplane bundles," Preprint, 1983.

[B7]. L. Boutet de Monvel. "Hypoelliptic operators with double characteristics and related pseudodifferential operators," Comm. Pure Appl. Math. 27(1974), 585-639.

[B8]. _____, "On the index of Toeplitz operators in several complex variables," Invent. Math. 50(1979), 249-272.

[B9]. _____, "Operateurs a coefficients polynomiaux, espace de Bargman et operateurs de Toeplitz," Sem. Goulaouic-Meyer-Schwartz 1980, Expose no. II bis.

[B10]. L. Boutet de Monvel, A. Grigis, and B. Helffer, "Parametrices d'operateurs pseudodifferentiels a caracteristiques multiple," Asterisque 34-35(1976), 93-121.

[B11]. L. Boutet de Monvel and V. Guillemin, The Spectral Theory of Toeplitz Operators, Princeton Univ. Press, 1981

[B12]. L. Boutet de Monvel and J. Sjöstrand, "Sur la singularité des noyaux de Bergman et de Szegö," Asterisque 34-36(1976), 123-164.

[B13]. L. Boutet de Monvel and F. Treves, "On a class of pseudodifferential operators with double characteristics," Comm. Pure Appl. Math. 27(1974), 59-89.

[C1]. A. Calderon and R. Vaillancourt, "A class of bounded pseudodifferential operators," Proc. Nat. Acad. Sci. USA 69(1972), 1185-1187.

[C2]. J. Chazarain, "Spectre d'un hamiltonien quantique et mecanique classique," Comm. PDE 5(1980), 595-644.

[C3]. M. Christ and D. Geller, "Singular integral characterization of Hardy spaces on homogeneous groups," Preprint, 1983.

[C4]. R. Coifman and G. Weiss, Analyze Harmonique Non-commutative sur Certains Espaces Homogenes, Lecture Notes in Math. 242, Springer, New York, 1971.

[C5]. Y. Colin de Verdiere, "Spectre conjoint d'operateurs qui commutent," Duke Math. J. 46(1979), 169-182.

[C6]. H. Cordes, "On compactness of commutators of multiplications and convolutions, and boundedness of pseudo-differential operators," J. Funct. Anal. 18(1975), 115-131.

[C7]. L. Corwin, "Tempered distributions on Heisenberg groups whose convolution with Schwartz class functions is Schwartz class," J. Funct. Anal. 44(1981), 328-347.

[C8]. L. Corwin and L. Rothschild, "Necessary conditions for local solvability of homogeneous left invariant differential operators on nilpotent Lie groups," Acta Math. 147(1981), 265-288.

[D1]. J. Duistermaat and V. Guillemin, "The spectrum of positive elliptic operators and periodic bicharacteristics," Invent. Math. 29(1975), 39-79.

[D2]. A. Dynin, "An algebra of pseudodifferential operators on the Heisenberg group; symbolic calculus," Soviet Math. Dokl. 17(1976), 508-12.

[D3]. _____, "Pseudodifferential operators on Heisenberg groups," Sem, CIME, Bressanone, Italy, June 1977.

[E1]. Yu. V. Egorov, "On canonical transformations of pseudo-differential operators," Uspehi Mat. Nauk. 24(1969), 235-236.

[F1]. C. Fefferman and D. Phong, "On positivity of pseudo-differential operators," Proc. Nat. Acad. Sci. USA 75(1978), 4673-4674.

[F2]. _____, "The uncertainty principle and sharp Gårding inequalities," Comm. Pure Appl. Math. 34(1981), 285-331.

[F3]. _____, "On the eigenvalue distribution of a pseudodifferential operator," Proc. Nat. Acad. Sci. USA 77(1980), 5622-5625.

[F4]. G. Folland and E. Stein, "Estimates for the $\bar{\partial}_b$ complex and analysis on the Heisenberg group," Comm. Pure Appl. Math. 27(1974), 429-522.

[F5]. _____, Hardy Spaces on Homogeneous Groups, Princeton Univ. Press, 1982.

[G1]. B. Gaveau, "Principe de moindre action, propagation de chaleur, et estimees sous elliptiques sur certains groups nilpotents," Acta Math. 139(1977), 95-153.

[G2]. D. Geller, "Local solvability and homogeneous distributions on the Heisenberg group," Comm. PDE 5(1980), 475-560.

[G3]. _____, "Fourier analysis on the Heisenberg group, I. Schwartz space," J. Funct. Anal. 36(1980), 205-254.

[G4]. _____, Ph.D. Thesis, Princeton Univ. 1977.

[G5]. _____, "Spherical harmonics, the Weyl transform, and the Fourier transform on the Heisenberg group," Preprint, 1983.

[G6]. D. Geller and E. Stein, "Estimates for singular convolution operators on the Heisenberg group," Preprint, 1983.

[G7]. R. Goodman, Nilpotent Lie Groups, Structure and Applications to Analysis, Springer Lecture Notes, no. 562, Springer, New York, 1976.

[G8]. P. Greiner, "On the Laguerre calculus of left-invariant convolution (pseudodifferential) operators on the Heisenberg group," Sem. Goulaouic-Meyer-Schwartz, 1980-81, expose no. XI.

[G9]. P. Greiner and E. Stein, Estimates for the $\bar{\partial}$-Neumann Problem. Math. Notes No. 19, Princeton Univ. Press, 1977.

[G10]. A. Grigis, "Hypoellipticite pour une class d'operateurs pseudo-differentiels a caracteristiques doubles et parametrix associes," C.R. Acad. Sci. Paris.

[G11]. A. Grossman, G. Loupias, and E. Stein, "An algebra of pseudodifferential operators and quantum mechanics in phase space," Ann. Inst. Fourier 18(1969), 343-368.

[G12]. V. Grusin, "Pseudo-differential operators on $\mathbb{R}^n$ with bounded symbols," Funct. Anal. and Appl. (1970).

[G13]. _____, "On a class of hypoelliptic operators," Math. USSR Sb. 12(1970), 458-476.

[G14]. V. Guillemin and S. Sternberg, Geometric Asymptotics, AMS, Providence, 1977.

[G15]. _____, "The metaplectic representation, Weyl operators, and spectral theory," J. Funct. Anal 42(1981), 128-225.

[H1]. B. Helffer, "Sur une classe d'operateurs hypoelliptiques a caracteristiques multiple," J. Math. Pure Appl. 55(1976), 207-215.

[H2]. _____, Theorie Spectrale Pour des Operateurs Globalement Elliptiques, Notas de Curso, Universidade Federal de Pernambuco. 1981.

[H3]. B. Helffer and J. Nourrigat, "Caracterization des operateurs hypoelliptiques homogenes invariant a gauche sur un groupe de Lie nilpotent gradue," Comm. PDE 4(1979), 899-958.

[H4]. B. Helffer and D. Robert, "Proprietés asymptotiques du spectre d'operateurs pseudodifferentiels sur $\mathbb{R}^n$," Comm. PDE.

[H5]. L. Hörmander, "Pseudodifferential operators and non-elliptic boundary problems," Ann. of Math. 83(1966), 129-209.

[H6]. _____, "Pseudodifferential operators and hypoelliptic equations," Proc. Symp. Pure Math. X (Singular Integrals), pp. 138-183. A.M.S., Providence, R. I., 1967.

[H7]. _____, "Fourier integral operators I," Acta Math. 127(1971), 79-183.

[H8]. _____, "A class of pseudodifferential operators with double characteristics," Math. Ann. 217(1975), 165-188.

[H9]. _____, "The Cauchy problem for differential equations with double characteristics," J. Anal. Math. 32(1977), 118-196.

[H10]. _____, "The Weyl calculus of pseudodifferential operators," Comm. Pure Appl. Math. 32(1979), 355-443.

[H11]. R. Howe, "Quantum mechanics and partial differential equations," J. Funct. Anal. 38(1980), 188-254.

[H12]. _____, "On the role of the Heisenberg group in harmonic analysis," Bull. A.M.S. 3(1980), 821-843.

[H13]. _____, "A symbolic calculus for nilpotent groups," Preprint.

[H14]. A. Hulanicki, "The distribution of energy in the Brownian motion in the Gaussian field and analytic-hypoellipticity of certain subelliptic operators on the Heisenberg group," Studia Math. 51(1976), 165-173.

[I1]. V. Ivrii and V. Petkov, "Necessary conditions for the Cauchy problem for non-strictly hyperbolic equations to be well-posed," Uspehi Mat. Nauk. 29(1974), 5.

[I2]. C. Iwasaki and N. Iwasaki, "Parametrix for a degenerate parabolic equation and its application to the asymptotic behavior of spectral functions for stationary problems," Publ. RIMS, Kyoto Univ. 17(1981), 577-655.

[K1]. T. Kato, Perturbation Theory for Linear Operators, Springer, New York, 1980.

[K2]. _____, "Boundedness of some pseudo-differential operators," Osaka J. Math. 13(1976), 1-9.

[K3]. N. Kerzman and E. Stein, "The Szegö kernel in terms of Cauchy-Fantappie kernels," Duke Math. J. 45(1978), 197-224.

[K4]. J. Kohn, "Harmonic integrals on strongly pseudoconvex manifolds," I and II, Ann. of Math. 78(1963), 112-147; 79(1964), 450-472.

[K5]. _____, "Pseudo-differential operators and hypoellipticity," Proc. Symp. Pure Math. 23, pp. 61-69, AMS, Providence, 1973.

[K6]. _____, "Microlocalization of CR structures," Abstracts AMS 4(Jan. 1983), p. 107.

[K7]. J. Kohn and L. Nirenberg, "An algebra of pseudo-differential operators," Comm. Pure Appl. Math. 18(1965), 269-305.

[K8]. A. Koranyi and S. Vagi, "Singular integrals in homogeneous spaces and some problems of classical analysis," Ann. Scuola Norm. Sup. Pisa 25(1971), 575-648.

[K9]. S. Krantz, Function Theory of Several Complex Variables, Wiley-Interscience, New York, 1982.

[K10]. H. Kumano-go, Pseudo-differential Operators, MIT Press, Cambridge, 1981.

[L1]. B. Lascar, "Propagation des singularities pour les equations hyperboliques a caracteristiques de multiplicite au plus double et singularities Masloviennes," Amer. J. Math. 104(1982), 227-285.

[L2]. B. Lascar and R. Lascar, "Propagation des singularities pour des equations hyperbolique a caracteristiques de multiplicite au plus double et singularities Masloviennes II," Jour. d'Anal. Math. 41(1982), 1-38.

[L3]. P. Lax and L. Nirenberg, "On stability for difference schemes; a sharp form of Garding's inequality," Comm. Pure Appl. Math. 19(1966), 473-492.

[L4]. N. Lebedev, Special Functions and Applications, Dover, New York, 1972.

[M1]. J. Mather, "Differential invariants," Topology 16(1977), 145-155.

[M2]. G. Mauceri, "The Weyl transform and bounded operators on $L^p(\mathbb{R}^n)$," J. Funct. Anal. 39(1980), 408-429.

[M3]. A. Melin, "Parametrix constructions for right invariant differential operators on nilpotent groups," Comm. PDE 6(1981), 1363-1405.

[M4]. _____, "Lie filtrations and pseudo-differential operators," Preprint, 1982.

[M5]. A. Melin and J. Sjöstrand, "Fourier integral operators with complex phase functions," Lecture Notes in Math. 459, pp. 120-223, Springer, New York, 1975.

[M6]. R. Melrose, "Hypoelliptic operators with characteristic variety of codimension two and the wave equation," Sem. Goulaouic-Schwartz 1979-80, Expose XI.

[M7]. _____, "The wave equation for a hypoelliptic operator with symplectic characteristics of codimension two," Unpublished manuscript, 1980.

[M8]. A. Menikoff and J. Sjöstrand, "On the eigenvalues of a class of hypoelliptic operators," Math. Ann. 235(1978), 55-85.

[M9]. G. Metivier, "Fonction spectrale et valeurs propres d'une classe d'operateurs non elliptiques," Comm. PDE 1(1976), 467-519.

[M10]. K. Miller, "Hypoellipticity on the Heisenberg group," J. Funct. Anal. 31(1979), 306-320.

[M11]. _____, "Parametrices for hypoelliptic operators on step two nilpotent Lie groups," Comm. PDE 5(1980), 1153-1184.

[M12] _____, "Invariant pseudodifferential operators on two step nilpotent Lie groups," Michigan Math. J. 29(1982), 315-328.

[N1]. A. Nachman, "The wave equation on the Heisenberg group," Ph.D. thesis, Princeton Univ., 1980.

[N2]. M. Nagase, "The L^p boundedness of pseudodifferential operators with non-regular symbols," Comm. PDE 2(1977), 1045-1061.

[N3]. A. Nagel and E. Stein, Lectures on Pseudo-differential Operators: Regularity Theorems and Applications to Non-elliptic Problems. Princeton Univ. Press, 1980.

[N4]. E. Nelson, "Operants, a functional calculus for non-commuting operators," in Functional Analysis and Related Topics, pp. 172-187. Springer, New York, 1970.

[N5]. L. Nirenberg, "On a problem of Hans Lewy," Uspehi Math. Nauk. 292(1974), 241-251.

[P1]. J. Peetre, "The Weyl transform and Laguerre polynomials," La Mathematische, Univ. de Catania Sem. 27, 301-323.

[P2]. D. Phong, "On integral representations for the Neumann operator," Proc. Nat. Acad. Sci. USA 76(1979), 1554-1558.

[P3]. D. Phong and E. Stein, "Some further classes of pseudo-differential and singular-integral operators arising in boundary-value problems I. Composition of operators," Amer. J. Math. 104(1982), 141-172.

[P4]. _____, "Estimates for the Bergman and Szegö projections on strongly pseudoconvex domains," Duke Math. J. 44(1977), 695-704.

[R1]. C. Rockland, Hypoellipticity and Eigenvalue Asymptotics, Lecture Note no. 464, Springer, New York, 1975.

[R2]. _____, "Hypoellipticity on the Heisenberg group: representation-theoretic criteria," Trans. AMS 240(1978), 1-52.

[R3]. L. Rothschild, "Local solvability of second order differential operators on nilpotent Lie groups," preprint.

[R4]. _____, "C^∞ regularity for solutions of invariant differential equations on the Heisenberg group," Preprint, 1982.

[R5]. L. Rothschild and E. Stein, "Hypoelliptic differential operators and nilpotent groups," Acta Math. 137(1976), 247-320.

[R6]. L. Rothschild and D. Tartakoff, "Inversion of analytic matrices and local solvability of some invariant differential operators on nilpotent Lie groups," Comm. PDE 6(1981), 625-650.

[R7]. W. Rudin, Function Theory on the Unit Ball of $\mathbb{C}^n$, Springer, New York, 1980.

[S1]. M. Sato, T. Kawai, and M. Kashiwara, "Microfunctions and pseudo-differential equations," in Hyperfunctions and Pseudo-Differential Equations, pp. 265-529. Lecture Notes in Math. 287, Springer, New York, 1973.

[S2]. A. Schwartz, "Smooth functions invariant under the action of a compact group," Topology 14(1975), 63-68.

[S3]. R. Seeley, "Complex powers of an elliptic operator," Proc. Symp. Pure Math., vol. X, pp. 288-307. AMS, Providence, R.I., 1967.

[S4]. I. Segal, "Transforms for operators and symplectic automorphisms over a locally compact abelian group," Math. Scand. 13(1963), 31-43.

[S5]. M. Shubin, Pseudodifferential Operators and Spectral Theory, Moscow, 1978 (In Russian).

[S6]. J. Sjöstrand, "Parametrices for pseudodifferential operators with multiple characteristics," Ark. för Math. 12(1974), 85-130.

[S7]. E. Stein, "An example on the Heisenberg group related to the Lewy operator," Invent. Math. 69(1982), 209-216.

[S8]. _____, Lectures on Pseudodifferential Operators, Course Notes, Princeton Univ., 1977.

[S9]. R. Strichartz, "A functional calculus for elliptic pseudodifferential operators," Amer. J. Math. 94(1972), 711-722.

[S10]. _____, "Invariant pseudo-differential operators on a Lie group," Ann. Scoula Norm. Sup. Pisa 26(1972), 587-611.

[T1]. D. Tartakoff, "Local analytic hypoellipticity of $\Box_b$ on nondegenerate Cauchy Riemann manifolds," Proc. Nat. Acad. Sci. USA 75(1978), 3027-3028.

[T2]. M. Taylor. Pseudodifferential Operators, Princeton Univ. Press, 1981.

[T3]. _____, "Fourier integral operators and harmonic analysis on compact manifolds," Proc. Symp. Pure Math., vol. 35, part 2, pp. 115-136, AMS, Providence, R.I., 1979.

[T4]. _____, "Pseudodifferential operators on contact manifolds, preliminary notes," Graduate lecture notes, Stony Brook, Spring 1981.

[T5]. _____, Basic Noncommutative Harmonic Analysis. Monograph, to appear.

[T6]. F. Treves, "Analytic hypoellipticity for a class of pseudodifferential operators," Comm. PDE 3(1978), 475-642.

[T7]. _____, Introduction to Pseudodifferential and Fourier Integral Operators, Two Volumes. Plenum, New York, 1980.

[U1]. A. Unterberger, "Oscillateur harmonique et operateurs pseudodifferentials," Ann. Inst. Fourier, Grenoble 29(1979), 201-221.

[U2]. _____, "Les operateurs metadifferentiels," in Complex Analysis, Microlocal Calculus, and Relativistic Quantum Theory, pp. 205-241, Lecture Notes in Physics, no. 126, Springer, New York, 1980.

[V1]. V. Varadarajan, Lie Groups, Lie Algebras, and their Representations, Prentice-Hall, Englewood Cliffs, New Jersey, 1974.

[V2]. M. Venugopalkrishna, "Fredholm operators associated with strongly pseudoconvex comains," J. Funct. Anal. 9(1972), 349-373.

[V3]. A. Voros, "An algebra of pseudodifferential operators and asymptotics of quantum mechanics," J. Funct. Anal. 29(1978), 104-132.

[W1]. N. Wallach, Symplectic Geometry and Fourier Analysis, Math. Sci. Press, Brookline, Mass. 1977.

[W2]. G. Warner, Harmonic Analysis on Semisimple Lie Groups, Springer, New York, 1972.

[W3]. A. Weil, "Sur certaines groupes d'operateurs unitaires," Acta Math. 111(1964), 143-211.

[W4]. H. Weyl, Group Theory and Quantum Mechanics, Dover, New York.

General instructions to authors for
PREPARING REPRODUCTION COPY FOR MEMOIRS

For more detailed instructions send for AMS booklet, "A Guide for Authors of Memoirs." Write to Editorial Offices, American Mathematical Society, P. O. Box 6248, Providence, R. I. 02940.

MEMOIRS are printed by photo-offset from camera copy fully prepared by the author. This means that, except for a reduction in size of 20 to 30%, the finished book will look exactly like the copy submitted. Thus the author will want to use a good quality typewriter with a new, medium-inked black ribbon, and submit clean copy on the appropriate model paper.

Model Paper, provided at no cost by the AMS, is paper marked with blue lines that confine the copy to the appropriate size. Author should specify, when ordering, whether typewriter to be used has PICA-size (10 characters to the inch) or ELITE-size type (12 characters to the inch).

Line Spacing – For best appearance, and economy, a typewriter equipped with a half-space ratchet – 12 notches to the inch – should be used. (This may be purchased and attached at small cost.) Three notches make the desired spacing, which is equivalent to 1-1/2 ordinary single spaces. Where copy has a great many subscripts and superscripts, however, double spacing should be used.

Special Characters may be filled in carefully freehand, using dense black ink, or INSTANT ("rub-on") LETTERING may be used. AMS has a sheet of several hundred most-used symbols and letters which may be purchased for $5.

Diagrams may be drawn in black ink either directly on the model sheet, or on a separate sheet and pasted with rubber cement into spaces left for them in the text. Ballpoint pen is *not* acceptable.

Page Headings (Running Heads) should be centered, in CAPITAL LETTERS (preferably), at the top of the page – just above the blue line and touching it.

LEFT-hand, EVEN-numbered pages should be headed with the AUTHOR'S NAME;

RIGHT-hand, ODD-numbered pages should be headed with the TITLE of the paper (in shortened form if necessary).

Exceptions: PAGE 1 and any other page that carries a display title require NO RUNNING HEADS.

Page Numbers should be at the top of the page, on the same line with the running heads.

LEFT-hand, EVEN numbers – flush with left margin;

RIGHT-hand, ODD numbers – flush with right margin.

Exceptions: PAGE 1 and any other page that carries a display title should have page number, centered below the text, on blue line provided.

FRONT MATTER PAGES should be numbered with Roman numerals (lower case), positioned below text in same manner as described above.

MEMOIRS FORMAT

It is suggested that the material be arranged in pages as indicated below.
Note: Starred items (*) are requirements of publication.

Front Matter (first pages in book, preceding main body of text).

Page i – *Title, *Author's name.

Page iii – Table of contents.

Page iv – *Abstract (at least 1 sentence and at most 300 words).

*1980 Mathematics Subject Classifications represent the primary and secondary subjects of the paper. For the classification scheme, see Annual Subject Indexes of MATHEMATICAL REVIEWS beginning in December 1978.

Key words and phrases, if desired. (A list which covers the content of the paper adequately enough to be useful for an information retrieval system.)

Page v, etc. – Preface, introduction, or any other matter not belonging in body of text.

Page 1 – Chapter Title (dropped 1 inch from top line, and centered).

Beginning of Text.

Footnotes: *Received by the editor date.

Support information – grants, credits, etc.

Last Page (at bottom) – Author's affiliation.

ABCDEFGHIJ–AMS–8987654